READING ABOUT SCIENCE
Skills and Concepts

BOOK G

John Mongillo
Beth Atwood
Kevin M. Carr
Linda J. Carr
Claudia Cornett
Jackie Harris
Josepha Sherman
Vivian Zwaik

Phoenix Learning Resources
St. Louis • New York

PHOTO CREDITS

ISBN 0-7915-2207-5

1 2 3 4 5 6 7 8 9 0 05 04 03 02 01 00

Authors

John Mongillo, Senior Author and General Editor
Science Writer and Editor
Saunderstown, Rhode Island

Beth S. Atwood
Writer and Reading Consultant
Durham, Connecticut

Kevin M. Carr
Teacher and Writer
Honolulu, Hawaii

Linda J. Carr
Writer and Psychologist
Honolulu, Hawaii

Claudia Cornett
Professor Emerita
Wittenberg University

Jackie Harris
Medical and Science Editor
Wethersfield, Connecticut

Josepha Sherman
Writer and Science Editor
Riverdale, New York

Vivian Zwaik
Writer and Educational Consultant
Wayne, New Jersey

CONTENTS

Page

Life Science 2

What Color Are Polar Bears? 6

Tiny Invaders...8

Dragons Alive! 10

Spiders Have Incredible Instincts.............................12

Helping Baby Fish Survive..14

Open Your Mouth, John Dory...................................16

Cattails: The Everything Plant 18

The Endangered Sea Cow 20

Strange Creatures Found in the Sea 22

What's for Lunch?..24

Underwater Geniuses 26

The Bee Team..28

Their Noses Know Not What They Smell 30

Insects Beware—A Dragonfly Is Near! 32

Heart Care Starts Early 34

At Home in a Coconut Shell 36

Meet the Mushroom 38

Life Under a Pier..40

Good Sports 42

Watch the Salt!...44

Earth-Space Science.. 46

Halley's Comet..50

Conquering the World's Peaks 52

Volcano Alert..54

Sea Secrets 56

The Mystery Planet 58

A Surprise Discovery 60

Eyes in the Skies..62

Litter in Space..64

Fascinating Lights: Auroras 66

The Woman, the Dog, and the Tent........................68

An Orbiting Telescope Peers into Space...................70

Page

The Weather Can Affect Your Life 72
Pluto and Charon: Two Celestial Bodies 74
Hidden Homes, Buried Buildings.............................76

Physical Science.. 78
Biking with Science on Your Side 82
Around the World in a Balloon!................................84
Does Your Car Have a Brain 86
Magic Waves...88
A Wonderful Tool: The Laser 90
Gliding Through Air 92
Tuna Surprise 94
Is There a Robot in Your Future?..............................96
Moving Heat Around 98
Trash: Something of Value 100
The Wonder of Natural Gas 102
Designing Cars is Not a Drag..................................104

Environmental Science... 106
China's Great Dam...110
What's Happening to the Ozone Layer? 112
Save the Wolf! 114
Green Sea Turtles Make a Comeback.........................116
Constant Energy From the Sun 118
Harnessing Steam from Volcanoes............................120
A Steady Welcome Wind 122
Clean Water Makes a Difference..............................124
Alternative Fuels 126
Getting Oil from Sand 128

Record Keeping 130

Metric Tables.. 132

Bibliography.. 134

Do you enjoy the world around you? Do you ever wonder why clouds have so many different shapes and what keeps planes up in the air? Did you ever want to explore a cave or find out why volcanoes erupt or why the earth shakes? If you can answer yes to any of these questions, then you will enjoy reading about science.

The world of science is a world of observing, exploring, predicting, reading, experimenting, testing, and recording. It is a world of trying and failing and trying again until, at last, you succeed. In the world of science, there is always some exciting discovery to be made and something new to explore.

Four Areas of Science

READING ABOUT SCIENCE explores four areas of science: life science, earth-space science, physical science, and environmental science. Each book in this series contains a unit on each of the four areas.

Life science is the study of living things. Life scientists explore the world of plants, animals, and humans. Their goal is to find out how living things depend upon each other for sur- vival and to observe how they live and interact in their environments, or sur- roundings.

Life science includes many special- ized areas, such as botany, zoology, and ecology. *Botanists* work mainly with plants. *Zoologists* work mostly with ani- mals. *Ecologists* are scientists who study the effects of air pollution, water pollution, and noise pollution on living things.

Earth-space science is the study of our Earth and other bodies in the solar sys- tem. Some earth-space scientists are *meteorologists*, who study climate and weather; *geologists*, who study the earth, the way it was formed, and its makeup, rocks and fossils, earth- quakes, and volcanoes; *oceanographers*, who study currents, waves, and life in the oceans of the world; and *astronomers*, who study the solar sys- tem, including the sun and other stars, moons, and planets.

Physical science is the study of matter and energy. *Physicists* are physical sci- entists who explore topics such as mat- ter, atoms, and nuclear energy. Other physical scientists study sound, mag- netism, heat, light, electricity, water, and air. *Chemists* develop the sub- stances used in medicine, clothing, food, and many other things.

Environmental science is the study of the forces and conditions that surround and influence all living and nonliving things. Environmental science involves all of the other sciences—life, earth-space, and physical.

If you want to know more about one or more of these areas of science, check the bibliography at the back of this book for suggested additional readings.

Steps to Follow

The suggestions that follow will help you use this book:

A. Study the photo or drawing that goes with the story. Read the title and the sentences that are printed in the sidebar next to each story. These are all clues to what the story is about.

B. Study the words for the story in the list of Words to Know at the beginning of each unit. You will find it easier to read the story if you understand the meanings of these words. Many times, you will find the meaning of the word right in the story.

When reading the story, look for clues to important words or ideas. Sometimes words or phrases are underlined. Pay special attention to these clues.

C. Read the story carefully. Think about what you are reading. Are any of the ideas in the story things that you have heard or read about before?

D. As you read, ask yourself questions. For example, "Why did the electricity go off?" "What caused the bears to turn green?" Many times, your questions are answered later in the story. Questioning helps you to understand what the author is saying. Asking questions also gets you ready for what comes next in the story.

E. Pay special attention to diagrams, charts, and other visual aids. They will often help you to understand the story better.

F. After you read the story slowly and carefully, you are ready to answer the questions on the questions page. If the book you have is part of a classroom set, you should write your answers in a special notebook or on paper that you can keep in a folder. Do not write in this book without your teacher's permission.

Put your name, the title of the story, and its page number on a sheet of paper. Read each question carefully. Record the question number and your answer on your answer paper.

The questions in this book check for the following kinds of comprehension, or understanding:

1. *Science vocabulary comprehension.* This kind of question asks you to remember the meaning of a word or term used in the story.

2. *Literal comprehension.* This kind of question asks you to remember certain facts that are given in the story. For example, the story might state that a snake was over 5 feet long. A literal question would ask you: "How long was the snake?"

3. *Interpretive comprehension.* This kind of question asks you to think about the story. To answer the question, you must decide what the author means, not what is said, or stated, in the story. For example, you may be asked what the main idea of the story is, what happened first, or what caused something to happen in the story.

4. *Applied comprehension.* This kind of question asks you to use what you have read to (1) solve a new problem, (2) interpret a chart or graph, or (3) put a certain topic under its correct heading, or category.

You should read each question carefully. You may go back to the story to help you find the answer. The questions are meant to help you learn how to read more carefully.

G. When you complete the questions page, turn it in to your teacher. Or, with your teacher's permission, check your answers against the answer key in the Teacher's Guide. If you made a mistake, find out what you did wrong. Practice answering that kind of question, and you will do better the next time.

H. Turn to the directions that tell you how to keep your Progress Charts. If you are not supposed to write in this book, you may make a copy of each chart to keep in your READING ABOUT SCIENCE folder or notebook. You may be surprised to see how well you can read science.

PRONUNCIATION GUIDE

Some words in this book may be unfamiliar to you and difficult for you to pronounce. These words are printed in italics. Then they are spelled according to the way they are said, or pronounced. This phonetic spelling appears in parentheses next to the words. The pronunciation guide below will help you say the words.

ă	pat	î	dear, deer, fierce, mere	p	pop	zh	garage, pleasure; vision
ā	aid, fey, pay			r	roar		
â	air, care, wear	j	judge	s	miss, sauce, see	ə	about, silent pencil, lemon, circus
ä	father	k	cat, kick, pique	sh	dish, ship		
b	bib	l	lid, needle	t	tight		
ch	church	m	am, man, mum	th	path, thin	ər	butter
d	deed	n	no, sudden	th	bathe, this		
ĕ	pet, pleasure	ng	thing	ŭ	cut, rough		
ē	be, bee, easy, leisure	ŏ	horrible, pot	û	circle, firm, heard, term, turn, urge, word		
		ō	go, hoarse, row, toe				
f	fast, fife, off, phase, rough	ô	alter, caught, for, paw	v	cave, valve, vine		STRESS
g	gag			w	with		Primary stress '
h	hat	oi	boy, noise, oil	y	yes		bi·ol'o·gy
hw	which	ou	cow, out	yōō	abuse, use		[bī ŏl'ejē]
ĭ	pit	ŏŏ	took	z	rose, size, xylophone, zebra		Secondary stress'
ī	by, guy, pie	ōō	boot, fruit				bi'o·log'i·cal
							[bī'elŏj'ĭkel]

LIFE SCIENCE

Deer have no permanent home. They spend their lives moving about within an area called their "home range." During the winter, they cluster in sheltered valleys, living on bark and other scarce vegetation. If there is not enough food to support the deer population, many of them starve. This is one reason a carefully controlled hunting season is permitted in many wilderness regions.

WORDS TO KNOW

What Color Are Polar Bears?
sufficient, enough
species, a kind, type, variety of creature
theory, an idea about how something came to be or about how something might be done, based upon some known information

Tiny Invaders
machinery, the inner workings
inflammation, redness, swelling
vaccination, a shot, or shots, with a vaccine (weakened or killed viruses or bacteria) to protect against a given disease

Dragons Alive!
forked, divided into two tips at the end
savage, fierce, vicious, brutal
tremendous, very great, large
extinct, no longer existing

Spiders Have Incredible Instincts
abdomen, belly
tarantulas, any of various large, hairy spiders

Helping Baby Fish Survive
predators, animals that hunt and eat other animals
incubator, a device for raising young living beings in the best conditions for healthy development

Open Your Mouth, John Dory
unique, one of a kind
legend, a popular story handed down from earlier times

The Endangered Sea Cow
endangered, in danger of no longer existing
mammal, a warm-blooded animal
aquatic, living or growing in or on the water
habitat, a place where a certain animal is likely to live

Strange Creatures Found in the Sea
exploratory, searching out
flexible, able to bend, twist

What's for Lunch?
nutritious, healthful, food that is good for a person
nutrient, the part of food used by our bodies for growth and energy
nutritionist, a food expert
dietitian, a person who knows proper nutrition and plans and supervises the preparation of meals

Underwater Geniuses
communicate, to exchange, or share, information
complicated, having parts, not easily understood

Their Noses Know Not What They Smell
phantom, not real
triggered, set into action
vibration, movement back and forth, a quiver

prey, animals hunted as food by other animals
pursuing, chasing

Insects Beware—A Dragonfly Is Near!
darting, moving very swiftly
beneficial, good for, helpful

Heart Care Starts Early
arteries, tubes carrying blood from the heart to all parts of the body
veins, any blood vessel bringing blood back to the heart
interfere, to hinder, prevent
circulatory system, the veins and arteries through which the heart pumps blood around the body to return to the heart and be pumped again
strains, to work too hard
hypertension, high blood pressure

At Home in a Coconut Shell
hermit, a person who chooses to live alone
scavengers, animals that eat dead or decaying plants and animals
thorax, the part of the body between the head and the abdomen (belly)
locomotion, power of moving from one place to another
cheliped, a leg or foot with a claw or pincers at the end
suitable, right for the purpose, fitting

Good Sports
incision, a cut into tissue or organ
therapy, treatment of disease or injuries
inflammations, redness, swelling

Watch the Salt!
vigorously, forcefully, powerfully

circulatory system, see Heart Care Starts Early above
heredity, all the characteristics that a person inherits from her or his parents

What Color Are Polar Bears?

What color are polar bears? White, of course! Why, then, were polar bears in a certain zoo turning green?

The keeper in the San Diego Zoo blinked her eyes. She could not believe what she was seeing. The polar bears had a greenish look! They were no longer white, as everyone knows polar bears should be. After this discovery, greenish-looking polar bears were also reported in several other zoos.

That is when scientists were called. The scientists discovered that the green color was caused by plants called *algae* (ăl′jē). Algae are simple plants found in fresh or salt water and in damp places on land. They use sunlight to make their own food and are an important food source for animals that live in the water.

When the scientists used microscopes to inspect the bears' fur, they found that the algae were actually growing inside the hollow shafts of the stiff hair on the bears' bodies. Even inside the hairs, the plants received sufficient sunlight to live and grow and give the bears' fur a greenish look.

The scientists examined the algae and discovered that they belonged to a common freshwater species that grows in lakes or swimming pools. The scientists had a theory that the algae in the bears' pools probably got into the hair shafts through breaks in the hair tips.

Although the algae did not harm the bears, the scientists thought it best to use a salt solution to kill the plants. Visitors to a zoo would expect to see white polar bears—certainly not green ones.

QUESTIONS

1. The word *algae* is used to describe simple _____ found in fresh or salt water and in damp places on land.

2. What did scientists use to inspect the fur on the bears' bodies?

3. The polar bears that were living in zoos were turning _____.

4. According to the story, in order to survive, algae must have
    a. sunlight.
    b. fresh water.
    c. salt.

5. The natural home of polar bears is the freezing cold Arctic. Their true color is useful because it
    a. camouflages, or hides, them from enemies.
    b. makes it easier for scientists to spot them.
    c. attracts algae, which help the bears' fur to grow.

6. The story says that the scientists had a theory about how the algae got into the bears' fur. The word *theory* means about the same as
    a. proof.   b. idea.   c. solution.

7. The algae found in the bears' fur belonged to _____ species.
    a. an unusual
    b. a harmful
    c. a common

8. For some reason, your green houseplants are wilting and dying. It could be that
    a. there is not enough salt in the plants' soil.
    b. the plants are not being watered properly.
    c. they are not located in a place that is dark enough for green plants.

Tiny Invaders

Are there ways to fight viruses?

Viruses are tiny organisms that enter your body, invade your cells, and can make you sick. Once within your cells, the virus takes control of the cells' machinery. Inside your cells, the virus makes many copies of itself. These new viruses leave the cell and move on to infect other cells. You get sicker.

Viruses cause flu, colds, inflammation of the brain and heart, and polio. You catch viruses from other people, animals, or even the bite of an insect.

The body has a few natural defenses against viruses. Only a vaccination offers any hope for preventing viral infections. While there are some antiviral medicines, the side effects often make the virus sufferer even sicker.

But now medical scientists are using a new method to develop antiviral medicines that have few, if any, side effects. Using powerful microscopes, they determine exactly what the virus looks like and how it works. Then the scientists make a medicine that locks onto the virus, blocking it at some point in its infection process. Two new anti-viral medicines, Relenza and Tamiflu, keep a virus from escaping from an invaded cell and infecting other cells. Another new medicine keeps the virus from making copies of itself.

The newest medicine, pleconaril, keeps the virus from invading the cell. This medicine, still being tested, appears to disable viruses that cause many serious infections including meningitis, an infection of the covering of the brain. And pleconaril pills may be the cure for the common cold.

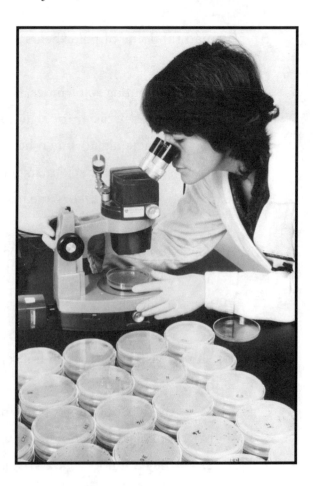

1. Tiny organisms that invade your cells are called

    a.   infections.

    b.   viruses.

    c.   anti-viral medicines.

2. Viruses make you sick when

    a.   you breathe them in.

    b.   someone sneezes.

    c.   they invade your cells.

3. Viruses make you sick because

    a.   they move from cell to cell.

    b.   they make copies of themselves.

    c.   you get a fever.

4. Select the event that might cause you to catch a virus.

    a.   You cut your hand.

    b.   Someone sneezes on you.

    c.   You drink sour milk.

5. Vaccination

    a.   doesn't prevent a virus infection.

    b.   cures the common cold.

    c.   prevents a virus infection.

6. The newest anti-viral medicine is

    a.   pleconaril.

    b.   Relenza.

    c.   Tamiflu.

7. Relenza and Tamiflu would be most likely to keep the flu from passing from person to person because they _____.

Dragons Alive!

Of all the known species of reptiles, the monitor lizard is the largest and heaviest species.

Reptiles (rĕp′ tīlz′) are cold-blooded animals with lungs and backbones. They are covered with scales, or horny plates, and usually lay eggs. One *species* (spē′shēz′), or type, of reptile, the monitor lizard, ranges in length from about 6 to 12 feet and can weigh as much as 250 pounds. Its forked tongue is long and slender, and it can turn its head in all directions.

The Komodo dragon is the largest of all known monitor lizards. Discovered in 1912 on the small island of Komodo in Indonesia, this lizard is extremely active and savage. It has saw-like teeth that can tear its prey—wild pigs, small deer, goats, and monkeys—into tiny, bite-sized pieces. The Komodo uses its powerful claws to dig holes for shelter. Its mighty tail is as long as its head and body put together. The lizard uses its tail as a weapon to fight off its enemies.

When the female Komodo lays her eggs in August, each one is about the size of a softball. Once they hatch, baby Komodos grow rapidly, and, by the age of five years they are fully grown. However, Komodos continue to gain weight because of their tremendous appetites.

The Komodo dragon is at home on land and in the water. During the day, it hunts in the rainforests and lowlands. On hot days, the lizard can cool itself in the water while it lazily digests a meal. At night, the Komodo sleeps in holes it has dug or among rocks or tree roots.

The Komodo dragon is considered to be a relative of the extinct giant lizards that roamed Earth before the existence of humans. For this reason, the government of Indonesia has passed laws protecting this living dragon.

1. The story describes _____ as cold-blooded animals with lungs and back bones. They are covered with scales, or horny plates.

2. The Komodo dragon is the largest of all known _____ lizards.

3. Which of the following describes the Komodo dragon?

    a.   active and savage

    b.   harmless and lazy

    c.   huge and timid

4. The Komodo dragon seems to prefer eating

    a.   plants.   b.   meat.   c.   fish.

5. If you saw a Komodo dragon that was about 12 feet long, you could be fairly certain that the lizard

    a.   would continue to grow in length.

    b.   was about 3 years old.

    c.   was an adult

Use the chart to answer questions 6, 7, and 8.

THE KOMODO DRAGON

Physical Feature	Description	Use
Skin	Rough, scaly	To protect itself
Teeth	Sawlike	To tear prey into small pieces
Claws	Powerful	To dig holes for shelter
Tail	As long as head and body together	To defend itself against enemies

6. Which physical feature of the Komodo dragon could be compared to a steak knife used by humans?

7. Humans use shovels and bulldozers to build houses and apartment buildings for shelter. The Komodo dragon uses its _____.

8. Humans use their hands, but the Komodo dragon uses its tail to _____.

Spiders Have Incredible Instincts

What do horses, cats, and spiders have in common? They are animals.

Spiders are misunderstood animals. First, they are not insects, as most people think. All insects have six legs, but spiders have eight. Spiders have only two body parts—the head and the abdomen—while insects have a third part called the thorax.

Second, most spiders are not poisonous. There are only two kinds, or *species* (spē'shēz'), of poisonous spiders in the United States. These two species are the black widow and the brown recluse. Even most tarantulas are not poisonous.

There are 30,000 different species of spiders. Of those ranging in size from 1 1/4 inches to 2 1/2 inches, only about one-half spin webs. All spiders produce silk, but each species has its own special way to use the silk. The web spinners use the silk to weave webs where they capture their prey. Probably the most familiar web is the circular web of the garden spider.

Other spiders make silky nets to drop over their victims. Another species sends out a single sticky strand of silk like a fishing line. When a fly is caught in the silk, the spider hauls in its "fish." Then the spider usually bites the insect and injects poison into it to paralyze or kill it. The spider then sucks out the victim's body juices. Because the spider has no mouth parts for chewing, it cannot eat solid food.

Many spiders have very poor eyesight, but one species, the jumping spider, has eight eyes. Jumping spiders can see perfectly in every direction. These spiders are also called "tiger spiders," because they sneak up and leap on their victims the way a tiger does.

QUESTIONS

1. Different kinds of spiders belong to one of the two _____.

2. How many legs do spiders have?

    a.   three   b.   six   c.   eight

3. The brown recluse and the _____ are the two poisonous spiders found in the United States.

4. According to the story, web spinners use their silky webs to

    a.   sleep in.

    b.   catch food.

    c.   hide themselves.

5. The story leads you to believe that some spiders eat juices from

    a.   fish.   b.   people.   c.   flies.

6. If you found a spider among some plants, its web would appear

    a.   round.

    b.   long.

    c.   thick.

7. How would you rate the jumping spider's ability to see?

1 2 3	4 5 6 7	8 9 10
POOR	**GOOD**	**VERY GOOD**

    a.   2

    b.   6

    c.   10

8. Under which of the following headings would you list spiders?

    a.   Clever Predators

    b.   Headless Animals

    c.   Helpful Insects

Helping Baby Fish Survive

Now baby fish have a better chance for survival.

Most kinds of fish lay hundreds or thousands of eggs. But most of the newly hatched fish are eaten by predators and do not have a chance to become fully grown. Now, Jeffrey Marliave, a marine biologist at the Vancouver Public Aquarium in British Columbia, Canada, has invented a device that protects baby fish.

Marliave's invention is a kind of incubator (ĭn'kyə bā'tər) for newly hatched fish. An incubator is a device that is used to promote the growth and development of young living organisms. It does this by providing the organism with a protective environment.

The incubator looks something like a big tub with a cone-shaped screen on one side and a rudder on the opposite side. The incubator is filled with newly hatched fish and then dropped into a body of water. The rudder directs the screened side of the incubator so that it faces the tide currents in the ocean water. In this way, the seawater can enter through the screen and circulate through the incubator. Tiny bits of plants and animals contained in the water serve as food for the young fish. The incubator also protects the baby fish from predators until they grow to full size and can take care of themselves. Then the fish can be released directly into the water or can be raised to a larger size and sold as food.

Marliave's invention could be used to help restock fishing grounds with certain saltwater fish that are popular with fishers. One such food fish is the lingcod found in Puget Sound.

1. Marliave's invention that protects baby fish is a kind of _____.

2. According to the story, most baby fish are eaten _____.

3. Opposite the screen side of the incubator is a _____.

4. How do the baby fish get fed in the incubator?

    a. Food is dropped into the incubator.

    b. They eat tiny pieces of plants and animals in the seawater.

    c. The incubator opens so that they can catch their own food.

5. The incubator serves to protect the baby fish from predators because

    a. its cone shape scares predators away.

    b. predators cannot fit through the screen.

    c. fishers trap the predators in the incubator.

6. The screen acts most like

    a. a filter.   b. a door.   c. bait.

7. Under which of the following headings would you list Marliave's invention?

    a. A Watertight Machine for Raising Baby Fish

    b. A Jail for Predators of Baby Fish

    c. A Life-Saving Device for Baby Fish

8. How would you rate the chances for survival of most newly hatched fish in open water?

1 2 3	4 5 6 7	8 9 10
POOR	GOOD	EXCELLENT

    a. 2

    b. 6

    c. 10

Open Your Mouth, John Dory

Who is John Dory? A fish, of course.

In terms of appearance, the John Dory is a unique fish. In the center of its body is a large, round black spot that is surrounded by a yellow ring. According to one legend, this spot is a thumbprint left over from Saint Peter when he took a coin out of the fish's mouth. When full grown, the John Dory may reach a length of about 2 feet. There are many stories about how this fish got its name. According to one story, fishers along the Adriatic Sea in Europe named their catch *il janitore*. In English, this name sounds something like "Johnny Dory," which, when shortened, becomes "John Dory."

The John Dory lives in the shallow oceans off the coasts of southern Europe and Africa. It swims at depths where it can be caught by fishers. And since it is an edible and good-tasting fish, fishers like to bring up lots of John Dories in their nets.

The John Dory is a *predator* (prĕd'ə tər), that is, it lives by capturing and feeding on other animals. This predator has an unusual way of catching its prey. The John Dory has jaws that can

separate from each other. When the predator gets close enough to its prey, it drops its lower jaw. Its mouth is now shaped somewhat like a large tube. The John Dory then sucks the fish it is pursuing into its tube-shaped mouth and snaps its lower jaw back to its upper jaw. This done, il janitore swims off eating its prey as if nothing has happened.

1. An animal that captures and lives off other animals is called a _____.

2. About how long can a full-grown John Dory be? _____

3. Why do fishers like to catch John Dories?

    a. They can be used as bait.

    b. They are predators.

    c. They are good to eat.

4. How do you know when the John Dory is ready to capture its prey?

    a. It snarls and bites.

    b. It drops its lower jaw.

    c. It dives down into the sea.

5. As the John Dory eats its prey, its upper jaw is

    a. opened wide.

    b. held in place.

    c. dropped down.

6. What makes the John Dory unusual looking?

    a. its yellow eyes

    b. its large jaws

    c. its black spot

7. Which of the following might you compare to the way in which a John Dory uses its mouth?

    a. a person drinking with a straw

    b. a bird tearing up an insect

    c. a person using a fork and a spoon

8. Under which of the following headings would you list the John Dory?

    a. Deep-Water Fish

    b. Predatory Fish

    c. Inedible Fish

Cattails: The Everything Plant

Beauty, food, and energy—cattails have it all.

Most people tend to think of cattails as merely pretty. Clumps of these graceful, bushy spikes add to the natural beauty of marshes and swamps. They can be just the right touch in a flower arrangement.

But soon, cattails may be known as "the everything plant." This is what some people in Minnesota think. Minnesota has a lot of swamps and a lot of cattails growing wild. Scientists at the University of Minnesota have been testing cattails. They have found that the plants could be a rich source of energy and food. Cattails could also provide the raw material to make paper.

The bushy top of the cattail is one energy source. Scientists have found that the bushy tops can be pressed into small pellets or bricks, which can then be burned as fuel. In addition, chemical processes can change the tops into alcohol or methane—both fuels.

The best food source is in the cattails' rhizomes (rīzōmz'). Rhizomes are root-like stems that grow under or along the ground. Chemical testing of cattail rhizomes shows that they are rich in sugar and starch. The rhizomes can be used to make flour or animal food. The starch and sugar in the rhizomes can be made into alcohol. The leaves and stems of the cattail plant can be ground up and used to make paper.

So do the people of Minnesota plan to cover their state with cattails? Not really. But many scientists do think that large fields of cattails could supply energy for small towns. In fact, some scientists think that cattails could supply 7 percent of Minnesota's energy some day soon.

1. Root-like stems that grow under or along the ground are called _____.

2. The cattail plant can be found growing in _____.
   a. food. b. swamps. c. chemicals.

3. Which of the following is true according to the article?
   a. The state of Minnesota is covered with fields of cattails.
   b. Cattails may someday supply energy for small towns.
   c. By the year 2000, cattails will supply most of Minnesota's energy.

4. In order to use the cattail plant to make paper, it is necessary to
   a. change the tops of the plant into alcohol.
   b. grind up the leaves and stems.
   c. treat the rhizomes with a special chemical.

5. In the article, cattails are called "the everything plant." They probably got this name
   a. as a result of scientific experiments on the plants.
   b. because people were eating the plants.
   c. after they were found growing all over Minnesota.

6. Before the bushy tops of the cattail can be used as an energy source, they must first be
   a. burned.
   b. pressed into small pellets.
   c. ground up into a flour-like substance.

7. Based on this story, which of the following words does not describe cattails?
   a. useful b. ugly c. rare

8. Under which heading would you classify both methane and alcohol?
   a. Fuels b. Plants c. Starches

The Endangered Sea Cow

You may have, if you have ever been in the state of Florida. That's where most of the few remaining manatee live in The United States. The manatee is a sea mammal. It lives in warm coastal waters, where it can feed on aquatic tropical plants.

Manatee live in small family groups or in herds of fifteen to twenty. They stay in warm, shallow, slow-moving waters where they can find food, shelter, and fresh water to drink. The female manatee is called a *cow*.

Most grown manatee are about 10 feet long and weigh about a thousand pounds. They are gray-brown, with bodies that are full in the middle and narrow at both ends.

The two small flippers on a manatee's upper body are used for steering. They are also used to bring food to its mouth. Grown manatee eat about 10 percent of their body weight each day.

Newborn manatee are called *calves*. They are a little over 4 feet long and weigh about 70 pounds. Although they may nibble on plants, they live mostly on the milk of their mothers.

Manatee spend about six to eight hours a day eating. They spend from two to twelve hours resting and travel or play for the remainder of the day.

One of the games played by manatee is "Follow-The-Leader." Several of them swim in single file, all swimming, breathing, and diving at exactly the same time.

At one time, there were many thousands of manatee in Florida. Today, boats and habitat destruction have reduced their numbers to only about 25,000. That's why this gentle sea mammal is listed as an endangered species.

1. The manatee is called a sea *mammal* because it

    a. lives in water and nurses its young.

    b. lives in warm waters.

    c. eats tropical plants.

2. Manatee live in

    a. the Atlantic Ocean.

    b. the Pacific Ocean.

    c. warm coastal waters.

3. Baby manatee are called

    a. cows.   b.   herds.   c.   calves.

4. Manatee spend most of their time

    a. playing.   b.   resting.   c.   eating.

5. Many manatee live in Florida because

    a. Florida has warm, slow-moving coastal waters where they can find food and shelter.

    b. they can play "Follow-the-Leader" there.

    c. they are protected from the dangers of humans there.

6. Grown manatee eat

    a. about 5 percent of their body weight each day.

    b. about 10 percent of their body weight each day.

    c. about 10 percent of their body weight each week.

7. Manatee are great fun to watch because

    a. they nurse their young.

    b. they are very playful.

    c. they eat aquatic plants.

8. The manatee is listed as an endangered species in Florida because

    a. many of them are killed by boats.

    b. there are only about 25,000 left living there.

    c. their habitat is disappearing.

Strange Creatures Found in the Sea

In 1979, an unusual form of life was discovered in the Atlantic Ocean, near Venezuela.

Strange red worms have been found living in the sea at depths of more than 2 miles. The sea is very dark at those depths, and the warm water is heated by hot metals, such as zinc, iron, and copper, which bubble up from beneath the sea floor.

Scientists found the red worms by chance during exploratory trips made in the tiny submersible *Alvin*. According to biologists, these red worms are unlike other worms in several ways.

The red worms can grow to be as long as 6 to 9 feet. They live in long tubes that are as tough and flexible as strong plastic. Just as other sea animals make their own protective shells, the red worms make their tube "homes."

The worm feeds through the feathery end of its body. Its food is bacteria that it sifts from the water. The bacteria are nourishing because they feed on thick layers of minerals found on the ocean floor.

The body of the sea worm is red because its blood contains a large amount of hemoglobin (hē′mə glō′bĭn). Hemoglobin carries oxygen to cells in the body and is the same chemical that makes blood red. The worm takes in oxygen from the surrounding seawater. Like most living things, the worm would die without oxygen.

The scientists brought back several giant worms for more study. One thing they need now is a new name for the red worm. The scientists like *Vestimentifera*. What do you think?

1. Hemoglobin is a chemical that carries _____ to cells in the body.

2. How does the red worm get its food?

    a.   through a long tube

    b.   through the feathery end of its body

    c.   from the shells of other sea animals

3. The water in which the worms live is warmed by hot

    a.   chemicals.   b.   metals.   c.   bacteria.

4. You can tell from the story that the red worms found by the scientists

    a.   do not need sunlight to live.

    b.   are very similar to other worms.

    c.   have very little hemoglobin in their bodies.

5. How are the red worms and most other living things alike?

    a.   Both must eat chemicals in order to survive.

    b.   Both must have oxygen to live.

    c.   Both need protective shells or coverings.

6. The zinc, iron, and copper found bubbling up from beneath the sea floor would come under the heading of

    a.   Minerals.   b.   Metals.   c.   Chemicals.

7. If you tried to bend the tubes in which the worms live, the tubes would most likely

    a.   bend without breaking.

    b.   snap in two like long, hard sticks.

    c.   remain rigid and not bend in any direction.

8. Fill in the missing link in the food chain below.

| Ocean Minerals | → | | → | Red Worms |

What's for Lunch?

What is good food? That question is on the minds of many experts who study food to see if it is healthful. Another word for healthful is *nutritious* (nōō trĭsh'əs). These food experts, or nutritionists, help plan well-balanced school lunches.

People eat about one-third of their daily food requirements at lunch. The food we eat should help us grow and develop properly. School dietitians make sure the lunch choices provide the *nutrients* (nōō'trē ənts) necessary for a healthy, well-balanced meal. Nutrients are the parts of food that are used by our bodies for growth and energy. Proteins, fats, carbohydrates, vitamins, and minerals are nutrients.

Foods contain varying amounts of nutrients. Foods high in calories supply the body with energy. The body uses protein to build new cells. Vitamin D—found in milk products— keeps bones healthy. Niacin helps us to grow. Certain minerals, such as iron, move oxygen through the body.

Nutritionists use new information about health and foods to help us plan healthy meals. Nutritionists have learned that eating too much fat, sugar, or salt is not healthy. School dietitians keep this information in mind when planning lunch menus. But they know that even a well-balanced meal will not nourish the student who does not eat it. So school dietitians are adding nutrients to students' favorites. Favorites include pizza, tacos, and milkshakes.

Nutritionists think if students learn about nutrition they will eat more healthful foods. What do you think?

QUESTIONS

1. The word in the story that means the same as *nutritious* is _____.

2. According to the story, nutrients are substances that are found in _____.

3. To make sure that meals you eat are well balanced, it is important to

    a. eat a variety of foods.

    b. add up the number of calories in each meal.

    c. choose foods that are especially rich in protein.

4. The main idea of this story is that

    a. too much salt is not healthy.

    b. nutritionists plan well-balanced meals.

    c. what we eat has an important effect on our health.

Use the information from the story to answer questions 5, 6, 7, and 8. Place the number of the nutrient beside what it does.

Nutrient

5. Protein

6. Iron

7. Vitamin D

8. Niacin

What It Does

a. Helps our bodies grow _____

b. Keeps bones healthy _____

c. Moves oxygen through the body _____

d. Builds new cells _____

Underwater Geniuses

As far back as 300 B.C., humans have been fascinated by those playful sea mammals that can "speak."

The Greek scientist Aristotle was the first person to report that dolphins could speak. But it is only recently that humans have begun to study closely the habits of the dolphins.

Dolphins are considered to be highly intelligent animals with well-developed brains. Perhaps the dolphins' most fascinating characteristic is their use of sound to communicate with each other.

Dolphins employ two distinct sound systems: *echolocation* (ĕk′ō lō kā′shən)

and "dolphin language." Echolocation is the ability to use high-frequency sounds to locate living and nonliving things. Dolphins use echolocation by sending out clicking sounds that bounce off underwater objects and return as echoes. The echoes give the dolphins clues to the shape, size, distance, and other characteristics of these objects.

Although dolphins have no vocal cords, they can emit whistles, squeaks, and groans. This second sound system, called dolphin language, is more complex and less understood than echolocation. However, research shows that at least some sounds made by dolphins, such as cries for help, are a form of language. Some scientists believe that dolphins can communicate complicated information. It seems, for example, that they are able to warn other dolphins to avoid fish nets.

Scientists are eager to understand more about dolphin language so that they can communicate with trained dolphins. If this communication is possible, dolphins might be sent on rescue missions to aid humans as well as underwater life. And some people think dolphins will be willing and able to help.

1. The dolphin's ability to locate objects by using sound is called _____.

2. Who first reported that dolphins could speak?

3. Dolphins are considered to be most interesting because of their ability to

    a. play tricks.   b. hear human sounds.

    c. use sound to communicate.

4. Before a dolphin can understand and interpret a click, the sound must

    a.   come back in the form of an echo.

    b.   pass through a bar-shaped object.

    c.   be sent to other dolphins nearby.

5. A dolphin's ability to emit squeaks, whistles, and groans is considered very unusual because a dolphin

    a. is a sea mammal.   b. has no vocal cords.

    c. does not have a highly developed brain.

Use the table to answer questions 6, 7, and 8.

Name	Occupation	Special Interest
Kenneth S. Norris	Biologist	Echolocation intelligence
Sylvia D. Earle	Marine biologist	Marine animals
John Cunningham Lilly	Neurophysiologist	Dolphin language intelligence
Louis Millerman	Psychologist	Dolphin language memory

6. According to the table, which scientist studies sea animals other than the dolphin?

7. Which two scientists would be especially interested in understanding a dolphin's whistles, squeaks, and groans?

8. Kenneth S. Norris probably has a good understanding of how dolphins can

    a. speak without vocal cords.   b. distinguish between objects.

    c. help humans with work.

The Bee Team

It is not easy to fool a bee.

A honey bee colony may contain as many as 30,000 to 40,000 bees. They survive the winter by clustering together in a dense ball. The safety and well-being of the hive depend on the bees' efficient teamwork.

Behavioral (bĭ hāv′yər əl) scientists study how humans and animals behave. They believe that honeybees act in a very organized way. A bee may accidentally pollinate, or fertilize, flowers while gathering nectar, a sweet liquid used by bees to make honey. But behavioral scientists believe that the finding and gathering of nectar is an organized and skillful process.

Recently, studies have been done on the bee's ability to "map" a route and "observe" geography. While a scout bee hunts for flowers, it keeps track of the sun, shadows, and any outstanding land features. Once it has found new sources of nectar, the bee makes a beeline for home, flying directly to a landmark, such as a tree or a rock, near the hive. Upon returning, the bee communicates its discovery to other bees through a dance. The nectar gatherers are then able to fly directly to the flower. Somehow, the scout bee has described the route, or path, to the nectar.

Behavioral scientists have done one study in which a sugar syrup is placed near a hive. Scout bees quickly locate the syrup and "spread the word." But after a few minutes, the scientists move the sugar 25 percent farther away from the hive. Soon, the bees catch on to this pattern and fly directly to the site. Once again, honeybees prove that they cannot easily be fooled.

1. In the story, *pollinate* means the same as _____.

2. Behavioral scientists study the behavior of

    a. humans and flowers.

    b. animals and humans.

    c. bees only.

3. The safety and well-being of the hive depend on

    a. how many flowers are pollinated.

    b. the bees' working together as a team.

    c. where the bees wander.

4. In the story, the term *beeline* is used to describe how a bee

    a. observes the geography of a place.

    b. heads for a landmark near its hive.

    c. hunts for new patches of flowers.

5. Which of the following statements is true?

    a. Honeybees are not able to survive cold temperatures.

    b. Scout bees have the ability to map routes.

    c. Bees can be tricked by changing the location of the nectar.

Worker bees perform several jobs. To answer questions 6, 7, and 8, put the number of the bees' job description on the line in front of the similar job done by humans.

6. feeding newborn bees

7. protecting the hive from enemies

8. cleaning empty cells in the hive for reuse

a. _____ guards

b. _____ nurses

c. _____ house workers

Their Noses Know Not What They Smell

Although our sense of smell is 10,000 times more sensitive than any of our other senses, it is usually taken for granted.

Most of us enjoy the odor of freshly baked bread, hot buttered popcorn, or sizzling bacon. But to some people the odors of fresh bread, burning rubber, and pizza are all the same. These people suffer from a condition called *anosmia* (ăn ŏz' mē ə). They have lost their sense of smell.

A normal sense of smell can provide information about distant objects. We do not have to be in the kitchen to know that the toast is burning. And we don't have to see a puddle to know that a gas tank is leaking. But people with anosmia are not warned of such unseen dangers.

A condition called *phantosmia* (fănt ŏz'mēə) causes people to smell odors that are not there—phantom odors. Seated at dinner, such a person may be sickened by the odor of a bowl of soup. Others at the table think that the same soup smells delicious. But the person with phantosmia smells something very different and may become quite ill.

Smell is a chemical sense, which means that it is triggered by chemicals in the things we smell. Scientists have several theories to explain why this is so.

One theory is based on the vibration, or movement, of the atoms in the molecules of things we smell. For example, if the chemicals in different objects have the same vibrations, the objects give off the same odor. According to another theory, if the molecules in things we smell have the same shape, the things will give off the same odor.

At this time, neither theory has been proved. Who knows what the nose knows?

1. People who suffer from a condition called _____ have lost their sense of smell.

2. People who smell odors that are not there suffer from a condition called _____.

3. According to the story, the sense of smell is triggered by _____ in the things we smell.

4. It is not necessary to see something in order to smell it.
    a. True
    b. False
    c. The story does not say.

5. According to one theory in the story, if two different foods smell the same, then it is possible that
    a. their chemicals have the same vibrations.
    b. they were cooked in the same chemical substance.
    c. they contain the same chemical ingredients.

6. Besides smell, the sense of _____ also gives us information about distant objects.
    a. taste   b. touch   c. hearing

7. Besides smell, the sense of _____ is also a chemical sense.
    a. sight
    b. taste
    c. touch

8. Match each sense in Column I with the part of the body in which it is found in Column II.

Column I	Column II
1. touch	a. _____ nose
2. taste	b. _____ eyes
3. smell	c. _____ skin
4. hearing	d. _____ ears
5. sight	e. _____ tongue

Insects Beware—A Dragonfly Is Near!

Some 400 different kinds of dragonflies live on or near the water in the United States.

Dragonflies fly swiftly from place to place, sometimes darting so quickly that they are difficult to see. They have two sets of wings that range from 2 1/2 to 5 inches long.

A dragonfly is considered to be one of the most beneficial, or helpful, insects to have around because it is a predator. A predator is an animal that preys on, or kills, other animals. The dragonfly is beneficial to humans because it eats other insects that humans consider to be pests. To catch the insects, the dragonfly puts its legs together and curves them so that they form a "basket." Then it uses this basket to scoop up insects from the air.

Dragonflies have an interesting life history. They lay their eggs underwater. The young hatch in about three weeks and are called nymphs (nĭmfs). Nymphs have no wings. They molt (mōlt), or shed their skins, as often as 10 to 15 times as they continue to live underwater. To survive, they prey on small underwater animals. It takes from one to four years before the last nymph stage is reached, and the nymph crawls out of the water. Then it splits its skin, and a winged adult dragonfly comes out.

After its wings dry, the dragonfly begins its role as a beneficial predator, preying on such insects as mosquitoes and flies. Sometimes, the dragonfly will eat insects that are considered beneficial to humans. But that is because all insects look like food to the hungry dragonfly!

1. The newly hatched young of the dragonfly are called _____.

2. Dragonflies help humans by eating _____.

3. Where does the dragonfly lay its eggs?

   a. in the sand

   b. on leaves

   c. in the water

4. You can tell when the dragonfly is attacking an insect because its

   a. wings flap up and down.

   b. skin splits open.

   c. legs curve to form a basket.

5. Before the dragonfly can become a predator, it must first _____ its wings.

   a. shed

   b. dry

   c. open

6. Which of the following statements best expresses the main idea of the story?

   a. To humans, the dragonfly is a beneficial predator.

   b. The dragonfly is one of our most annoying pests.

   c. The life history of the dragonfly is very interesting.

7. How would you rate a mosquitoe's ability to escape the dragonfly?

1 2 3	4 5 6 7	8 9 10
FAIR	**VERY GOOD**	**EXCELLENT**

   a. 3   b. 7   c. 10

8. Under which of the following headings would you list the nymph?

   a. Choosy Eaters

   b. Skin Shedders

   c. Strong Flyers

Heart Care Starts Early

Establishing good health habits early in life may help prevent disease.

As a result of research, most doctors now believe that it is important to care for your heart when you are young, before a problem can develop into something serious. By guarding against heart disease, you will ensure that your heart can do its job properly. The job of the heart is to pump blood to all parts of the body. Blood is carried away from the heart through blood vessels called *arteries* (är′tə rēz) and is returned to the heart through blood vessels called *veins* (vānz).

Many things can strain your heart and result in heart disease. Heart disease includes any condition that can weaken the heart or interfere with the way this marvelous circulatory system works. The most common problem is called hardening of the arteries. It occurs when deposits of a fatty substance called *cholesterol* (kə lĕs′tə rôl′) collect slowly inside the walls of the arteries. As the cholesterol builds up, the openings of the arteries get narrow and the walls of the arteries begin to harden. The heart must work harder to move blood through the narrowed arteries.

Overweight is another common problem. Being overweight strains the heart because the heart must circulate blood to the extra body fat. When blood flows through the blood vessels with too much force, high blood pressure, or hypertension, can result. Hypertension causes stress and strains your heart. Doctors now also think that eating too much salt often causes hypertension.

So be good to your heart. Avoid beef, butter, cheese, and other foods that are high in cholesterol and fat content. Reduce your food intake and keep your weight down by eating sensibly and exercising regularly.

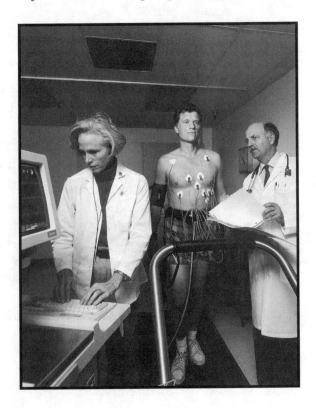

Use the terms below to answer questions 1 through 3.

veins   arteries   cholesterol   hypertension

1. A fatty substance present in many foods is called _____.

2. Blood travels to the heart through vessels called _____.

3. Too much salt in the diet may cause _____.

4. What effect does hardening of the arteries have on the heart?

     a. Blood flows to the heart with too much force.

     b. The heart works harder to pump blood.

     c. It is difficult for the heart to pump blood to extra body fat.

5. Which of the following statements is *true*?

     a. Overweight is the most common problem that leads to heart disease.

     b. Eating fatty foods may cause hardening of the arteries.

     c. Guarding against heart disease requires the help of a doctor.

6. Of the following, which would be the best food to avoid if you wanted to cut down on high-cholesterol foods?

     a. steak   b. chicken   c. fish

Use the diagram to answer questions 7 and 8.

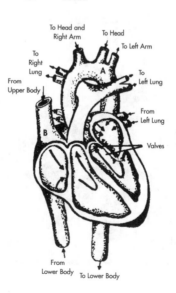

7. What type of blood vessel is labeled "A" in the diagram?

8. What type of blood vessel is labeled "B" in the diagram?

At Home in a Coconut Shell

Land hermit crabs have to be flexible when it comes to finding protection.

Hermit crabs live on land in shallow water or deep in the oceans, in both warm and cold-water regions. One kind of crab, the land hermit crab, lives in the Florida Keys.

The female land hermit crab carries her eggs inside her shell until they hatch. The young spend the first part of their lives in water. The rest of their lives is spent on land, but they always stay near water. Land hermit crabs are scavengers, which means that they eat dead or decaying material they find along the shore. Hermit crabs have become popular pets for people who enjoy unusual pets. Crab food is even available at some pet stores.

Hermit crabs have five pairs of legs and three main body parts—a head, a thorax, and an abdomen. Unlike some other crabs, the hermit crab's abdomen is not covered by a hard shell. So the hermit crab makes its home in empty shells, where it lives alone, like a hermit. It twists itself into a curved shell, leaving some of its legs sticking out for locomotion.

The crab's first pair of legs are *chelipeds*—meaning, they have claws. The right cheliped is larger than the left one. The chelipeds are used for defense and for capturing food. After the hermit crab withdraws into its shell, it uses its right cheliped to block the shell's entrance.

As the crab grows, it outgrows its protective shell and must look for a new home. The hermit crab sometimes kills a snail or another crab to steal its shell. If it cannot find a suitable new shell, the hermit crab may use things such as broken coconut shells and old plastic bottles for its home.

1. An animal that feeds on dead or decaying material is called a _____.

2. Where do land hermit crabs hatch their eggs?

    a.   in trees   b.   in water   c.   on land

3. Why is it necessary for hermit crabs to seek new homes?

    a.   They grow too large for their old shells.

    b.   They must change their hiding places.

    c.   They need new homes when they move to land.

4. Why is the crab in this story called a hermit?

    a. because it can climb trees   b. because it eats dead things

    c.   because it lives alone

5. Which of the following would be the best nickname for the hermit crab?

    a. The House Hunting Crab   b. The Unpopulars Crab

    c. The Bottle Crab

6. Which of the following would offer the most protection for a land hermit crab?

    a. a torn rag   b. an old shoe   c. a leaf

Use the diagram to answer questions 7 and 8.

7. To block the entrance to its shell, the hermit crab uses the part of its body that is labeled _____ on the diagram.

    a.   A   b.   B   c.   C

8. The first part of the hermit crab's body to enter the shell is called the

    a.   antenna.   b.   tail fan.   c.   abdomen.

Meet the Mushroom

What kind of plant does not have seeds, roots, or green leaves? It's the mushroom.

Mushrooms belong to a plant group called *fungi* (fŭn'jē). Mushrooms can be as beautiful and as colorful as wild flowers. Mushrooms do not have leaves or seeds and, unlike green plants, they cannot make their own food. Since they need a lot of moisture, mushrooms usually grow where it is damp. They may be found growing in decaying leaves, twigs, and stumps as well as in rich soil and on living trees.

The main part of the mushroom grows underground and is called the mycelium (mī sē'lē əm). The mycelium is the mass of thread-like, branching strands that forms the main part of a fungus. When the mushroom grows on a living tree, the mycelium causes the tree to decay. This decay is a source of food for the mushroom. The umbrella-like growth is a stalk that grows up from the mycelium. This stalk is really the fruit of the mushroom. The top of the "umbrella" is the cap. On the underside of the cap are the gills, from which the spores of the plant develop.

Mushrooms and other fungi make spores instead of seeds. When the fruit is ripe, the spores fall to the ground or are scattered by the wind. Some of the spores sprout, and a new mushroom plant develops. Spores can be white, yellow, pink, black, or brown. Their color helps identify those plants that are safe to eat. Only about 1,000 of the more than 38,000 kinds of wild mushrooms are safe to eat. The common table mush- room is grown in a special mushroom house and is safe and tasty to eat.

1. The mass of thread-like, branching strands that forms the main part of the mushroom is called the _____.

2. Mushrooms belong to a plant group called _____.

3. Unlike green plants, mushrooms cannot
    a. make their own food.
    b. grow in damp, wet places.
    c. be eaten.

4. According to the story, one way of distinguishing a safe mushroom from a poisonous one is by the
    a. size of its stalk.
    b. color of its spores.
    c. shape of its gills.

5. This story also might have been called
    a. "Growing Mushrooms."
    b. "All About the Mushroom Plant."
    c. "Plants That Are Safe to Eat."

Use the diagram below to answer questions 6, 7, and 8.

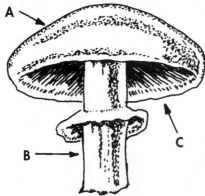

6. The parts of the mushroom from which the spores develop are labeled _____ and are called the _____.

7. The part of the mushroom called the cap is labeled _____.

8. The fruit of the mushroom is labeled _____.

Life Under a Pier

Living things are continually moving around beneath piers.

A *pier* (pîr) is a platform that extends from a shore out over water and can be used as a landing place for boats or ships. Most piers are supported by heavy wooden pillars, or *pilings* (pī'lĭngz), that have been sunk into the ocean floor.

There is a lot of activity going on around these underwater pilings. For example, it is not uncommon to find a piling covered with different kinds of sea animals, such as mussels, that live and feed at the water level. These shellfish attach themselves to pilings by tough "threads." By opening their shells to allow water to flow through, mussels are able to sift out tiny bits of plant and animal food from the water.

Starfish can also be found living near pilings, since they feed off shellfish, such as oysters and mussels, that have attached themselves to a piling. Starfish use their arms to pry open oyster and mussel shells. Then the starfish can easily extend their stomachs inside the shells to feed on the bodies of the shellfish.

Barnacles, anemones, sponges, and sea urchins also attach themselves to pilings. They, in turn, attract fish that swim in and around the pilings looking for a meal.

People often toss trash off piers, littering the ocean floor. In this litter living things are moving around, in search of food or a home. For example, octopuses can be seen swimming around. The octopus is a sea creature that makes good use of litter. For an octopus, "Home Sweet Home" may be an old tin can!

1. The wooden supports that hold up a pier are called pillars, or _____.

2. One function of a pier is to serve as a _____.

3. What do mussels use to attach themselves to piers?

4. What must a mussel do in order to obtain food?

    a. stick out its stomach

    b. open its shell

    c. attach itself to a plant or ships.

5. According to the story, what is the enemy of mussels and oysters?

    a. the sponge

    b. the sea urchin

    c. the starfish

6. Of the following, for what purpose might an octopus use litter?

    a. to make its home

    b. to sift out food

    c. to float around on

7. The activity that goes on around a piling results in a kind of _____.

    a. food chain

    b. hiding place

    c. shoreline

8. Fill in the following diagram:

    | | Mussels | Starfish |

    a. Barnacles

    b. Tiny bits of plants and animals

    c. Octopuses

Good Sports

Sports medicine takes care of athletes and tries to keep them from getting hurt. The first part of sports medicine is *prevention*. This means avoiding injuries by doing the right exercises. It also means practicing the right way so that athletes stay fit and strong. And it means avoiding using drugs such as anabolic steroids. All types of steroid drugs are related and are sometimes used to build muscle and bones. But if they aren't used properly, anabolic steroids can cause heart or liver disease or keep teenagers from growing to their proper height.

The second part of sports medicine is *nutrition*. This means making sure that athletes eat good, healthy meals and allow enough time between meals. This also means making sure athletes drink enough fluids, like water or fruit juice.

The third part of sports medicine is *treatment*. Sports accidents can happen even to professionals. When there is a smaller injury, like a tiny bone chip, a doctor can use orthroscopy, which is a way to look at a small injury without causing pain to the body. The method of removing the tiny bone chip through a very small incision is called orthoscopy. Wounds heal more quickly when there's not much surgery needed.

Careful exercising, called physical therapy, helps an injured athlete get back into shape. Physical therapists, experts in this type of sports medicine, decide what treatments to use for each athlete. Besides exercise, therapists may treat an athlete with whirlpool baths, which help soothe inflammations, or massages, which help an athlete's nervous system and keep muscles from getting weak.

Humans aren't the only athletes to be treated with sports medicine. Injured racehorses also may undergo orthoscopic surgery and physical therapy. They even have their own form of whirlpool baths.

1. The method of removing a bone chip through a very small incision is called _____.

2. The three parts of sports medicine are _____, _____, and _____.

3. According to the story, anabolic steroids if not used properly can cause

    a.   heart and liver disease.

    b.   heightened growth.

    c.   broken bones.

4. Physical therapists help athletes to

    a.   avoid injuries.

    b.   eat properly.

    c.   get the right treatment for injuries.

5. When an athlete needs surgery, orthoscopy, if possible, is the better way to operate because

    a.   doctors can see better using this method.

    b.   wounds heal more quickly.

    c.   it keeps the muscles from getting weak.

6. According to the story, which statement is true?

    a.   Humans are the only athletes that need sports medicine.

    b.   Humans aren't the only athletes that need sports medicine.

7. If an athlete needs help in recovering from an injury, the first person he or she should consult is

    a.   doctor.

    b.   a dietitian.

    c.   a physical therapist.

Watch the Salt!

We have all seen people use a salt shaker so vigorously that they practically cover the plate with salt. Sometimes they salt the food even before tasting it. But oversalting is not a good idea, because salt is thought to be dangerous for many people.

There is a lot of evidence that salt is a factor in causing high blood pressure, or hypertension. The sodium in table salt, or sodium chloride, holds water in the body. This makes it harder for a person's circulatory system to work properly and increases his or her blood pressure.

We do not know why some people get hypertension and others do not. One theory suggests that large amounts of salt taken in daily from childhood on can trigger hypertension in adults. Other possible causes of hypertension may include heredity, psychological factors, physical condition, and diet.

Although salt is essential, nutrition experts recommend that we control the amount of salt in our diet. Most of us could do very nicely with about 2 grams of salt a day. We can get this amount from our food without using additional salt. For example, a hot dog or a dill pickle has the same amount of sodium as 5 grams of salt.

About 30 million people in the United States are good candidates for high blood pressure. If someone in your family suffers from this condition, your own risk of getting hypertension is higher than normal. Think about watching *your* salt intake *now*.

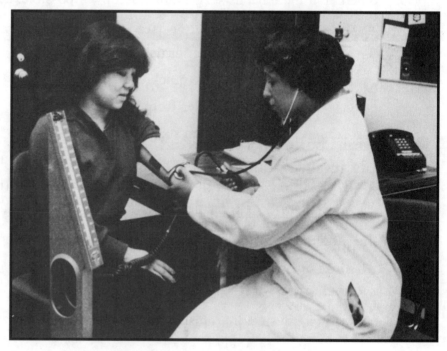

1. In the story, the word that means "high blood pressure" is _____.

2. Another term for table salt is _____.

3. Nutrition experts tell us that most people should _____ salt in their diets.

   a. increase the use of

   b. control the amount of

   c. stop using any

4. The story suggests that a low-sodium diet would be _____ for persons with high blood pressure.

   a. good   b. dangerous   c. useless

5. Today, doctors know how and why people get high blood pressure.

   a. True   b. False   c. The story does not say.

6. High blood pressure may begin at an early age. Therefore, it would be a good idea to

   a. have your blood pressure checked from childhood on.

   b. eat only between 5 and 6 grams of salt a day.

   c. stop eating any form of salt.

7. Fill in the missing link in the chain of events below.

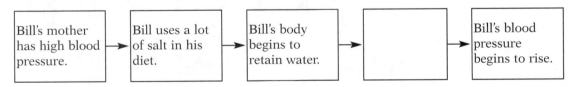

| Bill's mother has high blood pressure. | Bill uses a lot of salt in his diet. | Bill's body begins to retain water. | | Bill's blood pressure begins to rise. |

8. In the chain of events above, why would Bill's chances of having high blood pressure be greater than someone else's, even though the other person also uses a lot of salt?

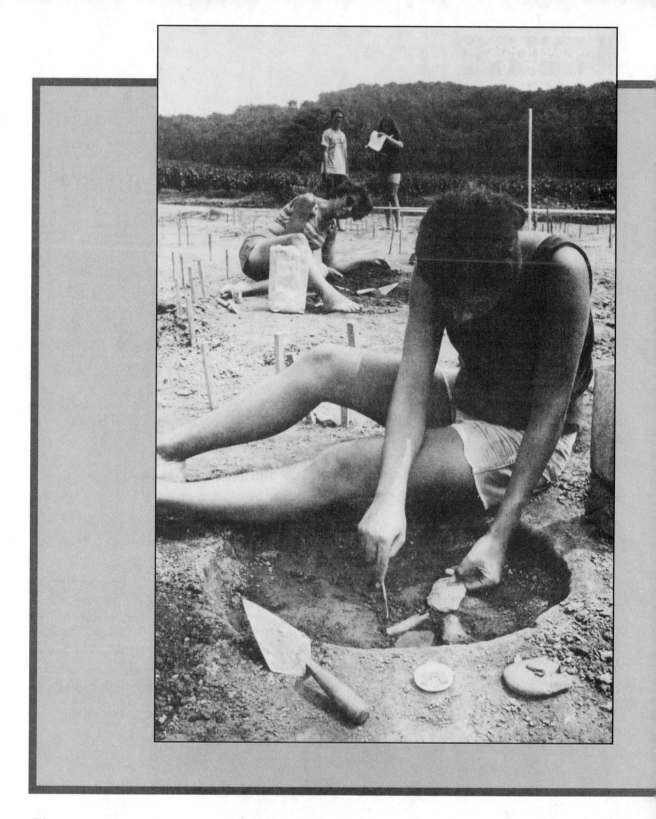

EARTH-SPACE SCIENCE

The Earth's crust is a storehouse of materials that are of interest to many different kinds of scientists. Paleontologists search for fossils to discover what kinds of plants and animals lived during various periods of the Earth's history. Archaeologists uncover ancient tools, buildings, and other objects to learn how people lived in the distant past. Mineralogists identify and study the minerals found in the Earth, and petroleum geologists explore for deposits of oil and gas.

WORDS TO KNOW

Halley's Comet
orbit, circular path around an object
nucleus, central part
particles, small parts

Conquering the World's Peaks
ascents, climbs

Volcano Alert
erupted, burst forth, broke out in a rush
spouted, shot out
violent, with great force

Sea Secrets
archaeologist, scientist who studies the life and culture of ancient peoples
exhausting, very tiring

The Mystery Planet
orbiting, circling
asteroids, the small planets that normally orbit the sun between Mars and Jupiter

A Surprise Discovery
fossils, the hardened remains of plants and animals
embedded, fixed firmly in

Eyes in the Skies
communication, the exchange of information
data, facts, figures, information

infrared, invisible light rays
monitor, to watch or check on

Litter in Space
rubbish, worthless stuff
litter, scattered rubbish

The Woman, the Dog, and the Tent
meteoroids, the small solid bodies traveling through outer space that are seen as meteors when they enter Earth's atmosphere

An Orbiting Telescope Peers into Space
optical, visual, for aiding vision
galaxies, vast groupings of stars

The Weather Can Affect Your Life
behavior, the way a person acts, behaves
dilate, to make wider or larger, expand
constrict, to make smaller or narrower

Pluto and Charon: Two Celestial Bodies
celestial, of the heavens
interplanetary, within the solar system, between planets

Hidden Homes, Buried Buildings
subterranean, underground
maintain, to continue, keep up

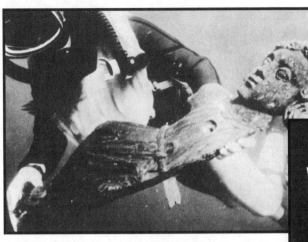

Halley's Comet

Bright comets are spectacular sights, but they are not often seen by the human eye.

In 1682, the English astronomer Edmund Halley identified and studied the motion of a bright comet. First, he figured out the comet's orbit (ôr′bĭt), or path, around the sun. Then Halley studied the records of two bright comets observed in 1531 and 1607. He decided that all three were actually the same comet returning to Earth's view about every 76 years. Based on these studies, Halley predicted that the comet would reappear in 1758. His prediction was right, and on December 24, 1758, the comet returned. Since then, every 76 years the comet, now called Halley's comet, comes close enough to Earth to be seen either through a telescope or by the naked eye.

The main part of a comet is its nucleus (nōo′klē əs). The nucleus is made up of frozen water, methane, and other gases. Billions of tiny pieces of rock are frozen into the "ice." As you can see in the diagram on the next page, a comet can be very near the sun at one point in its orbit and very far away at another point. As the comet comes closer to the sun, the icy material of its nucleus starts to melt and evaporate. A tail of gas and rock particles forms and may stream millions of miles into space.

Solar winds cause the bright, glowing tail always to point away from the sun. A comet's tail is longest when it is nearest the sun. As it moves farther away from the sun, the tail gets smaller and smaller until it finally disappears.

When it appeared in 1910, Halley's comet could be seen clearly in the sky. When the comet appeared in 1985/86, it was farther from the Earth. Light pollution from Earth dimmed some of the comet's light. But it could still be seen without a telescope.

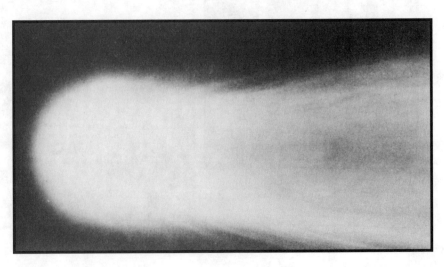

Use the word list below to answer questions 1, 2, and 3.

nucleus   orbit   telescope

1. A comet's path around the sun is also called its _____.

2. The main part of a comet is its _____.

3. An instrument that makes faraway objects easier to see is called a _____.

4. The first recorded sighting of Halley's comet was in the year _____.

5. What causes a comet's tail to form?

    a. solar winds   b. solar heat   c. solar ice

6. After its appearance in 1986, Halley's comet will return again in the

    year _____

The orbit of Halley's comet is shown in green in the diagram below. *Use the diagram to answer questions 7 and 8.*

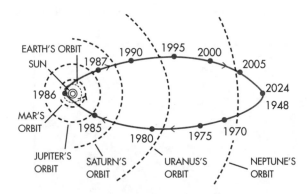

7. The year before it approached Earth's orbit in 1986, Halley's comet was just outside the orbit of

    a. Saturn.   b. Jupiter.   c. Mars.

8. When will Halley's comet be in the same place as it was in 1948?

Conquering the World's Peaks

During the 1700s, scientists began asking questions about the structure and origin of mountains. To find the answers to these questions, they *ascended* (ə sĕnd′ĕd), or climbed, some of the world's most interesting mountains.

Many of the early ascents were made in the French, Swiss, and Italian Alps. The Swiss scientist De Saussure made one of the first successful climbs of Mont Blanc, a Swiss mountain. While on the mountain, De Saussure studied glaciers, rock formations, and high-altitude weather patterns. News of De Saussure's climb set off a wave of interest in mountain climbing throughout Europe. Soon, not only other scientists but pure adventurers began to climb mountains.

From the late 1700s through the 1800s, many famous mountains were climbed for the first time. Among these were the Matterhorn in Switzerland, Kilimanjaro in Africa, Mount Whitney in California, and Mount McKinley in Alaska. The middle 1800s became known as the "Golden Age of Mountain Climbing." During this time, the first mountain-climbing clubs were formed.

Climbers soon turned their attention to the most challenging mountain range in the world, the *Himalayas* (hĭm′ə lă′əz). This mountain range, on the continent of Asia, includes the highest peak on Earth, Mount Everest. Everest reaches 29,028 feet into the air and was not ascended successfully until 1953, when Sir Edmund Hillary set foot on the summit. It has been climbed a number of times since then.

1. In the story, the word _____ means "climbed."

2. The Golden Age of Mountain Climbing occurred during the middle

    a.   1700s.   b.   1800s.   c.   1900s.

3. One of the first people to climb Mont Blanc successfully was _____.

4. Besides climbing for scientific reasons, people also climbed mountains for adventure.

    a.   True   b.   False   c.   The story does not say.

5. Sir Edmund Hillary made history by being the first person to

    a.   form a mountain-climbing club.

    b.   climb to the top of Mount Everest.

    c.   study rock formations and glaciers.

Use the table to answer questions 6, 7, and 8.

HIGHEST AND LOWEST ALTITUDES ON THE SEVEN CONTINENTS

Continent	Highest Point	Feet Above Sea Level	Lowest Point	Feet Below Sea Level
Asia	Mount Everest, Nepal-Tibet	29,028	Dead Sea, Israel-Jordan	1,290
South America	Aconcagua, Argentina	22,605	Valdes Peninsula, Argentina	129
North America	Mount McKinley, Alaska	20,117	Death Valley, California	281
Europe	Mount Elbrus, U.S.S.R.	18,325	Caspian Sea, Russia	502
Antarctica	Vinson Massif	16,191	—	—
Australia	Mount Kosciusko, New South Wales	7,237	Lake Eyre, South Australia	53

6. The lowest mountain listed is on the continent of _____.

7. Mount McKinley is ranked the _____ highest peak in the world.

    a.   second   b.   third   c.   fourth

8. Which of the following is the lowest point in the world?

    a.   Lake Eyre   b.   Death Valley   c.   the Dead Sea

Volcano Alert

On June 9, 1991 the volcano Pinatubo on the island of Luzon in the Philippines erupted. It had been sleeping for 600 years.

The eruption sounded like thunder rolling across the island. Molten rock, or lava, and hot gases spouted to a height of 24 miles from this opening in the Earth's crust. A gray, cauliflower-shaped cloud 11 miles wide rained ash and cinders.

Seven days later, this most violent and destructive volcanic event of the twentieth century was over. But for 10 to 15 years after the event, mudflows caused by heavy rains might sweep down river valleys. They might wash out roads and villages. These mudflows might also bury low-lying areas in several yards of mud and volcanic ash.

Mount Pinatubo is called an *active* volcano because it has erupted at least once in the past 10,000 years. There are about 1,500 active volcanoes in the world.

A *dormant* volcano is one that has not erupted in 10,000 years, but which is expected to erupt again. Extinct volcanoes are those that have not erupted for 10,000 years. But now and then, a supposedly extinct volcano, such as Lassen Peak in California, suddenly erupts.

Volcanologists, scientists who study volcanoes, use sensitive instruments to help them predict volcanic eruptions. Some volcanoes give warnings before they erupt. The ground near them begins to shake slightly. The delicate instruments feel and record the shaking.

When the volcanologists receive this information, they know that an eruption is about to take place. Then people living near the volcano can be warned. Because of these warnings, the lives of thousands of people were saved when the great Pinatubo erupted in 1991.

1. A volcano that has erupted within the last 10,000 years is said to be _____.

2. According to the article, a volcano is actually an opening in the Earth's _____.

3. When a volcano erupts, molten rock, or _____, and hot _____ pour out of the opening.

4. The instruments that scientists use to predict an eruption must be very sensitive to

    a.   movement.   b.   sound.   c.   temperature.

5. A dormant volcano will probably never erupt again.

    a.   True   b.   False   c.   The story doesn't say.

6. A volcano that has not erupted within recorded history is said to be

    a.   active.   b.   extinct.   c.   dormant.

Use the chart below to answer questions 7 and 8.

Volcano	Location	Date of Eruption	Results
Krakatoa	Indonesia	1883	Over one-half of the island of Krakatoa was destroyed completely; great sea waves drowned 36,000 people on nearby islands; the explosion was felt more than 1,600 kilometers away.
Mount Pelee	St. Pierre, Martinique (Caribbean Island)	1902	There was a fiery explosion; gases and dust poured from the crater, smothering 30,000 people in St. Pierre and leaving 1 survivor.
Mount Usu	Japan	1977	Hot ashes poured from the crater; 20,000 tourists and 7,000 residents fled nearby towns; there was serious crop damage.

7. The most violent eruption shown on the chart was located in _____.

8. Which volcano caused people to be smothered by gas and dust?

Sea Secrets

Are you hunting for the remains of an ancient city or a cargo of treasure? Try looking underwater.

In the 1600s, a Portuguese ship sank off the coast of Kenya. One of the objects recovered from the shipwreck can be seen in the picture below. As you can see, the 300-year-old wooden angel appears to be in fairly good condition.

Burial at sea slows down or even stops the normal processes of decay. Wood, cloth, and other fibers rot very slowly underwater as compared with above ground. Unlike things above ground, objects underwater are not subjected to the effects of wind, rain, and extreme changes in temperature.

Until recently, underwater exploration was an archaeologist's nightmare. But scuba-diving techniques and equipment now make it possible for divers to move around more freely, and wet suits protect them in cold waters. The development of underwater lights, cameras, and other equipment has also made the exploration easier.

However, underwater exploration is exhausting and often dangerous. It is also more time-consuming than working on land *digs*. (A *dig* is the term used to describe the place where archaeologists are working.) Divers usually work in pairs or small groups, and they can remain on the ocean bottom for only short periods of time. Then they must return to the surface very slowly because of the change in water pressure as they rise. Rising too quickly could be fatal.

But the historical treasures raised from the deep are worth the effort. Archaeologists have found "drowned" villages from prehistoric times, flooded cities, and wrecked treasure ships. At last, the sea is giving up some of its secrets.

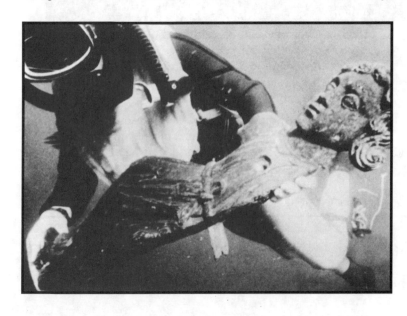

1. Places where archaeologists are working are known as _____.

2. Special techniques and equipment allow divers to move around _____ underwater.

3. Underwater, the normal processes of decay

    a.   are slower than they are above ground.

    b.   destroy only wood and cloth objects.

    c.   do not take place.

4. According to the story, what type of equipment enables divers to explore in cold-water areas?

5. Which of the following statements is *not* true of underwater exploration?

    a.   Searches require a lot of time.

    b.   Archaeologists work in small teams.

    c.   Divers can work for long periods of time.

6. In recent years, techniques and equipment for underwater archaeology have been greatly

    a.   improved.

    b.   reduced.

    c.   neglected.

7. Scientists who dig up the remains of underwater treasures might be called _____ archaeologists.

    a.   field   b.   marine   c.   lunar

8. How would you rate the risks involved for a diver exploring the ocean floor?

1 2 3	4 5 6 7	8 9 10
LOW	AVERAGE	HIGH

    a.   1   b.   6   c.   9

The Mystery Planet

Which two planets in our solar system are thought of as twins?

The planets Venus and Earth are roughly the same size. That is why they have been called the twins of our solar system. Venus also used to be known as the "mystery planet." Because dense, or thick, layers of clouds always cover the surface of Venus completely, astronomers were not able to see it with their telescopes.

Then, late in 1978, the *Pioneer 1* spacecraft began orbiting Venus. *Radar* (rā'där') was used to penetrate the thick layers of clouds. Radar is special equipment that uses radio waves to locate distant objects and determine what they are like. Maps were made from *Pioneer 1*'s findings. These maps show that the largest canyon in our solar system is on Venus. The canyon is about 900 miles long and about 168 miles wide. It is about 3 miles deep.

The map also indicated a ridge of mountains. The mountain ridge stands on a broad plateau that scientists think may be the result of earthquake movements. In the ridge is a volcano called Maxwell. It is about 12,210 yards high. Mount Everest, the highest mountain on Earth, is about 9,570 yards high.

The *Magellan* spacecraft orbiting Venus in the early 1990s produced a complete map of the planet's mountainous and cratered surface. Comets and asteroids hitting the planet caused the craters. One very large crater, Cleopatra Patera, is more than one and one-half miles deep. The *Magellan* also found Venus's atmosphere to be made of layers of oxygen and carbon dioxide—similar to Earth's atmosphere.

But the surface of Venus is very different from the surface of Earth. Scientists have learned that Venus is very hot and dry. The surface temperature on Venus is about 455°C (850 °F). Slowly, Venus is becoming less of a mystery.

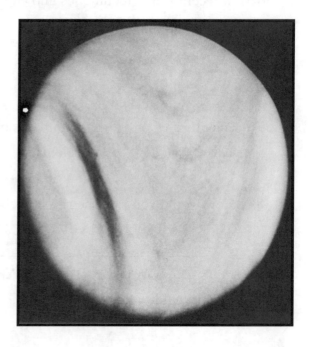

1. Special equipment that uses radio waves to locate distant objects and determine what they are like is called _____.

2. Up until 1978, astronomers were unable to see the surface of Venus because of its thick _____ layers.

3. Compared with Mount Everest, Maxwell is
    a. broader. b. higher. c. deeper.

4. Craters on Venus were caused by
    a. earthquakes. b. the impact of comets. c. gravity.

5. Which of the following describes the surface of Venus?
    a. high and low b. smooth and flat c. gaseous and cold

Use the table to answer questions 6, 7, and 8.

SOLAR-SYSTEM STATISTICS FOR FIVE OF THE NINE PLANETS

Planet	Average Distance from Sun (in miles)	Diameter of Planet at Its Center (in miles)	Number of Years to Revolve Around Sun
Mercury	34,740,000	2,880	0.24
Venus	64,920,000	7,260	0.62
Earth	89,760,000	7,654	1.00
Mars	136,800,000	4,070	1.88
Jupiter	466,800,000	85,920	11.86

6. Compared with other planets, the planet that is farthest from the sun takes the _____ amount of time to revolve around the sun.
    a. least b. most c. same

7. It takes Venus _____ to revolve around the sun.
    a. more than 6 years b. less than 1 year c. about 12 years

8. According to the table, the planet with the second largest diameter is
    a. Mars. b. Jupiter. c. Earth.

A Surprise Discovery

In 1975, an important discovery was made at the bottom of an old gold mine in central Alaska. The mine is located near the mining ghost town of Jack Wade in the Yukon-Tanana uplands. However, it was not gold that was discovered in the mine, but the fossilized bones of animals nearly 30,000 years old. *Fossils* (fŏs′əlz) are the remains of ancient plant and animal life. The Jack Wade fossils were of extinct animals, that is, animals that no longer live on Earth, such as bison, musk oxen, and wooly mammoths. Lee Porter found these animal bones while earning her doctorate degree in geology.

Why was this discovery so important? In addition to dating animal activity in the area, the fossils provided *archaeologists* (är′kē ŏl′ə jĭsts) with clues to human activity. Archaeologists study the remains of past human activities. The bones had been deeply embedded in silt at the bottom of the gold mine. This means that the animals must have lived and died right near Jack Wade and were not washed ashore from somewhere else. The bones were scarred and had burn marks on them. So Porter believes that during the Ice Age in North America, early humans may have decided to use the bones as tools.

The Jack Wade fossils are one of the earliest known records of human activity in North America. The former site of the 1898 Gold Rush has turned into an archaeologists' gold mine. Lee Porter's discovery has shown how humans in search of prey, as well as shelter from advancing Ice Age glaciers, worked their way across a land bridge from Siberia to the Alaskan Yukon.

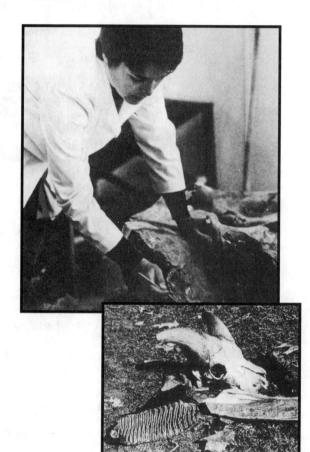

1. The word in the story that means "the remains of ancient plant and animal life" is _____.

2. The bones that were found were from _____ animals.

3. Where were the animal bones found?

4. The animals must have lived and died near Jack Wade because the bones had been
   a.   washed ashore nearby.   b.   deeply embedded in silt.
                        c.   found right there.

5. How did Porter know that humans had also been near Jack Wade 30,000 years ago?
   a.   The animal bones showed signs of human use.
   b.   Humans had left behind a written record.
   c.   Human bones were found with the animal bones.

6. How would you rate Porter's discovery?

1 2 3	4 5 6 7	8 9 10
NOT IMPORTANT	IMPORTANT	VERY IMPORTANT

   a.   2
   b.   5
   c.   9

Use facts from the story and the map to answer questions 7 and 8.

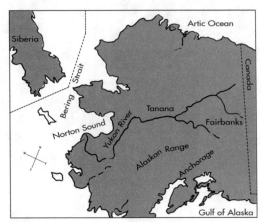

7. From which direction were early humans traveling?
   a.   east to west   b.   north to south   c.   west to east

8. What body of water did the Siberia-to-Alaska land bridge cross?
   a.   the Yukon River   b.   the Bering Strait   c.   Norton Sound

Eyes in the Skies

Today, most weather data is collected by machine.

During World War I, meteorologists (mē'tē ə rŏl'ə jĭsts), or weather scientists, did not receive reports on weather in other countries. Because of the war, communication among countries in Europe was very difficult. So a team of meteorologists in Norway experimented with new ways to predict the weather. They developed a theory that polar fronts, or masses of cold air, move down from the Arctic and affect weather around the world. Forecasting aids such as weather balloons, aerial photographs, and weather planes did not exist. So the meteorologists collected data, or information, supporting this polar-front theory from telephone and telegraph reports and fishers returning to shore. Then this information was collected and analyzed.

Today, radar, television, infrared photography, and weather satellites inform us of distant weather conditions instantly. In minutes, computers analyze and report data that it would take weeks to process by hand. For example, radar pictures provide meteorologists with data on the moisture content, structure, and location of clouds. Weather satellites orbiting Earth send back signals of the world's cloud cover. Meteorologists turn the signals into pictures. They monitor, or watch, these photographs and use their knowledge of clouds to predict the weather.

Satellites monitor temperatures above Florida citrus groves. When frost is probable, growers are warned in time to heat their groves. Meteorologists are using the information collected by satellites and other modern machinery to develop new theories about the weather.

1. A polar front is a mass of _____.

2. Weather scientists are also called _____.

3. According to the story, during World War I, weather information was collected and analyzed by _____.

4. According to the story, one important way in which computers help meteorologists is by saving them _____.

5. The Norwegians contributed to the history of weather forecasting by

    a. inventing the use of infrared cameras.

    b. developing a theory of air movement.

    c. sending up the first weather satellite.

Use the chart below to answer questions 6, 7, and 8.

MAIN TYPES OF CLOUDS AND THE
LEVELS AT WHICH THEY OCCUR

Cloud Family	Average Height Range (in feet)	Type of Cloud	Symbol
The High Clouds	20,000 to 40,000	Cirrus Cirrostratus Cirrocumulus	Ci Cs Cc
The Middle Clouds	6,500 to 20,000	Altostratus Altocumulus	As Ac
The Low Clouds	1,600 to 2,000	Stratus Nimbostratus Stratocumulus	St Ns Sc
The "Vertical" Clouds	1,600 to 40,000+	Cumulonimbus Cumulus	Cb Cu

6. To which cloud family do clouds ranging in height from 6,500 to 20,000 feet belong?

7. Which cloud types have the greatest range in height?

8. Which symbols stand for the Low Clouds?

Litter in Space

> *Even a small piece of metal or plastic can be a big problem for space flights.*

A satellite is any object that orbits, or circles, another one. For example, the moon is a natural satellite that orbits the Earth. In 1957, humans launched the first artificial, or man-made, satellite into orbit around the Earth. It was very small, and because it was launched into a low orbit, it fell back to Earth after 92 days. Since that first man-made satellite, we have sent thousands of objects into higher orbits around our planet. These objects remain in space for much longer periods of time. Many of these objects serve useful purposes, but some are now just rubbish.

This litter in space, or "orbital debris," as it is commonly referred to, comes mainly from exploded rockets and satellites that did not perform as they were expected to. Millions of pieces of metal and plastic are traveling around the Earth at speeds that average 22,000 miles per hour. Most of these pieces of debris are very small. Because of the high speed at which they travel, even a very small object might cause serious damage to a spacecraft.

To combat this problem, the Space Surveillance Network was created. *Surveillance* means to keep a close watch over someone or something. The Space Surveillance Network (SSN) uses information gathered from telescopes and electronic tracking stations around the world to keep a close watch over orbital debris. The smallest object the SSN can track, or follow, is about 4 inches in diameter. Computers help keep track of thousands of objects that might pose a threat to space traffic. This information is shared with everyone, to help make space flights as safe as possible.

1. Orbital debris comes mainly from
    a. telescopes and electronic tracking stations.
    b. the moon.   c. exploded rockets and satellites.
2. A satellite is
    a. any object that circles another one, and it can be natural or artificial.
    b. a computer on the ground used to track orbital debris.
    c. the Space Surveillance Network.
3. Orbital debris is thought to be dangerous because
    a. the larger pieces could hit the moon.
    b. it might damage spacecraft.
    c. it is litter.
4. Which of the following pieces of space litter would probably fall back to Earth first?
    a. A rocket that exploded in a low orbit
    b. A satellite that stopped working in a high orbit
    c. A spacecraft that ran out of fuel as it neared Mars
5. How many pieces of orbital debris are believed to be in orbit around the Earth?
    a. hundreds   b. thousands   c. millions
6. The information the Space Surveillance Network has on their computers would be useful to you if you were planning a space mission because
    a. it could help you decide how much fuel your rocket should carry.
    b. it could show you where to find spare parts in space.
    c. it could warn you of areas in space you should avoid.
7. The Space Surveillance Network has a difficult job because
    a. there are just a few pieces of space debris in orbit.
    b. there are millions of pieces of debris to keep track of.
    c. there aren't enough satellites to keep track of all the debris.
8. Assuming an increasing number of space launches, indicate whether you think each statement is true or false.
    a. The Space Surveillance Network will probably have more objects to track in the near future.   TRUE   FALSE
    b. If the amount of space debris increases, the risk to spacecraft will go down.   TRUE   FALSE
    c. Space mission planners should pay close attention to orbital debris.   TRUE   FALSE

Fascinating Lights: Auroras

Huge curtains of green and red light move slowly across the night sky. What causes these fabulous light displays?

These special, many-colored lights seen in the night skies are known as *auroras*, (ô rôr′əz), after the Roman goddess of dawn, Aurora. We call the lights in the Northern Hemisphere the *aurora borealis* (bôr′ ē ăl′ĭs). The lights in the Southern Hemisphere are called the *aurora australis* (ô strā′ lĭs).

Both auroras occur when electrically charged particles from the sun hit and react with the Earth's atmosphere. This collision causes the molecules to glow like fluorescent lights.

Auroras occur most frequently when there is a lot of sunspot activity. Sunspots are large, dark areas on the sun where the gases are "cooler" than in the surrounding areas. Some astronomers think that sunspots are storms on the sun. These storms send tremendous streams of charged particles toward Earth.

Although auroras can occur in nearly every part of the sky, they are usually too faint to be seen except in the polar regions of the Northern and Southern Hemispheres. The aurora borealis, or the northern lights, is visible in the northern part of North America. Green and red are the colors that occur most often there, but it is also possible to see white and yellow.

At first, the lights appear low on the horizon, but when they reach full power, they may cover the entire sky. The auroras may occur in streamers and arcs of light as well as in sheets, or curtains. The constantly changing display of the auroras is one of nature's most fascinating shows.

1. The special, many-colored lights seen in the Northern Hemisphere are called _____.

2. The special, many-colored lights seen in the Southern Hemisphere are called _____.

3. The colors that occur most often in the Northern Hemisphere are _____ and _____.

4. Auroras occur only in the skies near the Earth's polar regions.

    a.   True   b.   False   c.   The story does not say.

5. Sunspots are dark because they are

    a.   next to bright-colored auroras.

    b.   surrounded by hotter gases.

    c.   actually storms on the sun's surface.

6. After reading the story, you would conclude that people living near the equator most likely

    a.   see auroras only at dawn.

    b.   never see auroras.

    c.   see red lights, but not white or yellow lights.

7. Fill in the missing link in the chain of events below. Select a, b, or c.

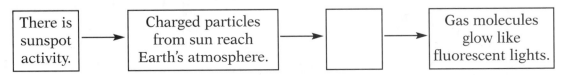

    a.   Storms occur on the sun and give off molecules of gas.

    b.   Fluorescent lights hit charged particles.

    c.   Charged particles react with The Earth's atmosphere.

8. Look at the diagram below. Which letter on the diagram indicates where auroras first appear?

    a.   A

    b.   B

    c.   C

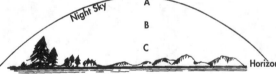

The Woman, the Dog, and the Tent

After 73 years, three meteorites, are reunited.

Meteoroids (mē′ tē ə roidz′) are stone or metal particles that revolve around the sun like very tiny planets. When part of a meteoroid falls to Earth, it is called a meteorite. The largest meteorite displayed anywhere in the world rests in the Hall of Meteorites at the American Museum of Natural History in New York City. Known as Ahnighito, or the Tent, the meteorite is composed mostly of nickel and iron and weighs over 75,000 pounds. The name *Ahnighito* comes from an Innuit legend, or story, that tells of a woman, her dog, and a tent that fell from the sky.

The Tent and two other meteorites—the Woman and the Dog—were found in Greenland in the late 1800s. You can see Ahnighito, the Tent, in the photograph, which was taken at the site where the meteorites were found. In 1906, Ahnighito and the 6,600-pound Woman were brought to the Museum of Natural History and placed near an entrance to the building. The museum's Hayden Planetar-ium was built around the meteorites in 1935.

Then, in 1979, both meteorites were removed from the planetarium through a huge hole that had been cut in a wall of the building. They were taken by trailer to a parking lot, where they remained under guard for seven days. During that time, Ahnighito was hoisted into the air by a 180-foot crane. Its underside was measured, and six points were marked to show where six sturdy columns would support the meteorite in its new home.

At the end of the week, Ahnighito and the Woman were taken by truck to the museum's Hall of Meteorites. There, they were reunited with their 1-ton companion, the Dog, from which they had been separated for 73 years.

1. Stone or metal particles that revolve around the sun like very tiny planets are called _____.

2. When were the meteorites in this story found?

3. According to the story, the name *Ahnighito* comes from an _____ legend.

4. Which fact mentioned in the story gives a clue to the size of the meteorites?

    a. They were guarded for seven days.

    b. The Hayden Planetarium was built around them.

    c. They now rest in the Hall of Meteorites.

5. After reading the story, you could conclude that

    a. meteorites can cause damage.

    b. most meteorites are made of iron and nickel.

    c. meteorites are very rarely found in Greenland.

6. What was the importance of the six points on Ahnighito?

    a. They were original markings found on Ahnighito.

    b. They told the legend of the meteorites.

    c. They indicated where Ahnighito needed support.

7. People who wanted to see the meteorites in 1968 went to

    a. the Hall of Meteorites.

    b. the Hayden Planetarium.

    c. a lot near the Museum of Natural History.

8. Ahnighito would be classified as a

    a. nickel-and-iron meteoroid.

    b. stone meteorite.

    c. metal particle from space.

An Orbiting Telescope Peers into Space

How big is the universe? How old is it? How did it begin? The Hubbel Space Telescope may provide some answers.

In the 1990s, NASA launched the space shuttle that carried the Hubbel Space Telescope. Astronauts placed it in orbit 300 miles above the Earth. The orbiting space telescope was the largest optical instrument ever placed in space. Optical instruments help us to see objects more clearly.

The *telescope* (tĕl′ ə skōp′) uses mirrors to gather visible light, which makes observation of distant objects possible. Even objects that are too faint to be seen by the eye are brilliantly visible in this telescope.

The telescope is used to study space objects such as stars, galaxies, planets, and comets. Light from these objects enter the open end of the telescope and is then projected by a large mirror onto a smaller mirror.

From there, the light is directed toward scientific instruments in the back of the telescope. These instruments take pictures and measure the distance of space objects as well as determine their makeup. The information is beamed back to astronomers at Earth stations.

So far Hubbel has allowed astronomers to see thousands of galaxies. The space telescope provided information for astronomers to make the first surface map of Pluto. Astronomers observed the collision between Jupiter and parts of a comet. The space telescope also was used to study the atmospheres of the planets and dust storms on Mars.

1. A telescope gathers _____ light and makes it possible to observe distant objects.

2. The Space Telescope orbits the Earth at a height of _____ miles.

3. To gather visible light, a telescope uses _____.

4. Poor viewing of space objects from telescopes on Earth is caused by
    a.   clouds.   b.   distance.   c.   darkness.

5. After light passes onto the smaller mirror of the space telescope, it
    a.   gets projected onto a larger mirror.
    b.   is directed toward instruments in the back of the telescope.
    c.   passes out the open end of the telescope.

6. Which of the following is not a function of the scientific instruments?
    a.   to measure the distance of space objects
    b.   to photograph Earth from space
    c.   to discover what space objects are made of

7. Which of the following would not be an optical instrument?
    a.   a microscope
    b.   a pair of glasses
    c.   a television

8. How would you rate the ability of the space telescope to view the universe?

1 2 3	4 5 6 7	8 9 10
POOR	GOOD	VERY GOOD

    a.   3
    b.   6
    c.   10

The Weather Can Affect Your Life

Can weather affect the way you feel and act?

There is a growing belief among behavioral scientists that, indeed, weather does have an effect on our daily lives. According to Dr. Stephen Rosen, an expert on weather and human behavior, there are many ways in which weather affects us. As Dr. Rosen explains it, weather is a form of stress, and its changes will result in stressful effects on the body.

Although our bodies adapt to changes in the weather, we are not always prepared for sudden weather changes and extremes in temperature. For example, in warm weather, blood vessels *dilate* (dī lāt'), or expand, to allow the body to get rid of excess heat. In cold weather, the blood vessels constrict, or tighten, keeping warmth in the body. Either change in the blood vessels triggers other bodily changes, for example, in body chemistry, blood composition, and the amount of oxygen going to the brain. It is this last change that most directly affects our mood and behavior. With a sudden change in the weather, we may feel happy and full of energy one day, depressed and run-down the next.

Extremes in temperature also affect people who take medication. For example, when you take aspirin, it causes the blood vessels to dilate, which in turn causes the body to lose heat more quickly. Other types of drugs or medication will affect the body differently according to the type of weather.

So it is important for people to be aware of the effects of weather and take medication under a doctor's supervision. Keeping in touch with your body's needs may mean tuning in to tonight's weather forecast.

1. In the story, the word _____ means "to expand."

2. Dr. Rosen describes weather as a form of _____.

3. What happens to a person's blood vessels in cold weather?

4. Unless you are under a doctor's orders, you probably should not take too many aspirins in _____ weather.

    a.   hot   b.   damp   c.   cold

5. Which of the following statements *best* expresses the main idea of the story?

    a.   Weather forecasts can predict your health.

    b.   Changes in the weather affect people's lives.

    c.   Medication should never be taken in excess.

Use the table below to answer questions 6, 7, and 8.

TIPS FOR WARM AND COLD CLIMATES

	Warm or Hot Weather	Cold Weather
Eat	salads, vegetables, and carbohydrates such as breads and cereals.	meats, cheese, milk, eggs, and other foods high in protein and fat.
Drink	lots of water and other fluids; replace water hour by hour.	normal amounts of liquid.
Avoid	emotional stress, long periods of heavy work outdoors.	too much dry heat, long periods of heavy work outdoors.

6. During hot weather, your body _____ liquids.

    a.   retains   b.   loses   c.   stores

7. In cold climates, you are advised to eat foods rich in _____ and _____.

8. What should you avoid in both hot and cold weather?

Pluto and Charon: Two Celestial Bodies

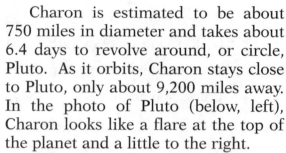

Pluto and its moon were discovered in much the same way.

On June 22, 1978, James W. Christy was working at the U.S. Naval Observatory examining some photographic plates of the planet Pluto. He looked at the photographs that had been taken by telescope in Flagstaff, Arizona. Christy soon realized that he had discovered a new *celestial* (səlĕs′ chəl) body, a moon of Pluto! The word *celestial* is used to describe something related to the sky or the heavens. This new moon was given the identification number 1978-P-1. Christy suggested the name *Charon* (kâr′ ən).

Charon is estimated to be about 750 miles in diameter and takes about 6.4 days to revolve around, or circle, Pluto. As it orbits, Charon stays close to Pluto, only about 9,200 miles away. In the photo of Pluto (below, left), Charon looks like a flare at the top of the planet and a little to the right.

Pluto was discovered in much the same way. Dr. Percival Lowell believed there was another planet in the universe. He had worked out a mathematical formula indicating the planet's supposed location. Many photographs of this celestial area were taken with a telescope. Then, in 1930, 14 years after Lowell's death, Clyde W. Tombaugh was working in Flagstaff. Looking through photographic plates, he discovered an object, a new celestial body. Tombaugh had sighted the outermost planet in the solar system, Pluto.

Some astronomers think that once Pluto may have been a moon of the planet Neptune and that it was knocked out of orbit during an interplanetary "accident."

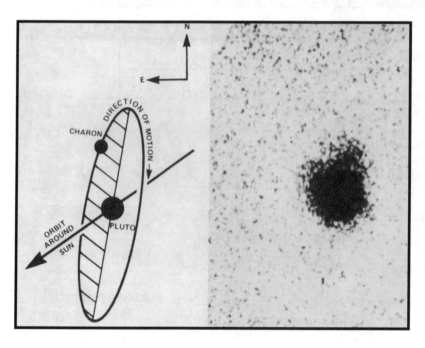

1. The word used to describe something related to the sky or the heavens is _____.

2. How was Pluto's moon first identified?

    a.   by name   b.   by number   c.   by location

3. According to the story, some scientists believe that at one time Pluto may have been a moon of _____.

4. Which scientist was responsible for estimating where Pluto was located in space?

5. The discoveries of Pluto and Charon were similar in that both bodies

    a.   were discovered by the same person.

    b.   had been photographed by telescope.

    c.   were sighted in the same year.

6. Pluto is about 1,560 miles in diameter. According to information in the story, Charon's diameter would be

    a.   less than half that of Pluto.   b.   twice that of Pluto.

                      c.   about the same as Pluto's.

Use the table to answer questions 7 and 8.

The Planets in Order of Distance from the Sun	Period of Revolution Around the Sun
Mercury (the closest)	88 days
Venus	225 days
Earth	365 days
Mars	687 days
Jupiter	12 years
Saturn	29 1/2 years
Uranus	84 years
Neptune	164 years
Pluto (the farthest)	247 years

7. The planet that takes the longest time to revolve around the sun is the

    a.   farthest away from the sun.   b.   fourth planet from the sun.

                      c.   closest planet to the sun.

8. Compared with Earth, it takes Uranus _____ more years to revolve around the sun.

Hidden Homes, Buried Buildings

Today, architects are building down under instead of up above.

Rock caves and dirt-bank cellars are examples of *subterranean* (sŭb′tə rā′nē ən), or below-ground-level, structures. They may sound like cold, uncomfortable places, but both caves and subsoil maintain almost constant temperatures. In New England, the air temperature yards below the ground is 50°F, regardless of the temperature of the air above ground. The air temperature at the entrance to a cave can be 95°F, but a hundred or so yards inside, the air temperature is a steady 52°F, summer or winter. In Kansas City, Missouri, over 100 companies employing 2,000 people have subterranean offices in limestone caverns.

"Earth-sheltered" structures are also being built. Unlike the subterranean type, earth-sheltered structures are not truly underground. They might be built into a hillside or halfway below ground level. For example, an elementary school in Virginia was built into a hillside, and only its solar collectors and skylight are visible. It costs half as much to heat and cool this school as it does similar schools in the same area.

"Aren't earth-sheltered homes damp and cold and slimy and dark?"

people ask. No, they don't have to be. Skylights and windows, set into the slope of the earth, let in light and sunshine.

There are some drawbacks to building down, however. The condition of the soil must be right. Hard rock or moist, swampy areas would not be good building locations. Also, it is fairly difficult and costly to add on to subterranean and earth-sheltered structures.

1. In the story, the word that refers to structures built below ground level is _____.

2. The air temperature in caves and subsoil _____ as the air temperature outside rises.

    a. rises slightly   b. remains steady   c. falls sharply

3. Living in a cave is a dark and gloomy existence.

    a. True   b. False   c. The story does not say.

4. Which of the following is not true of earth-sheltered buildings?

    a. They are cheaper to heat and cool.

    b. They maintain constant year-round temperatures.

    c. They can be built anywhere.

5. During which season would the air temperature of New England subsoil be most useful to people living in earth-sheltered homes?

    a. summer   b. fall   c. winter

Use the table to answer questions 6, 7, and 8.

EARTH-SHELTERED AND SUBTERRANEAN BUILDINGS

Place	Terrain or Setting	Structure	Natural Temperature
Waxahachine, Texas	10 feet below ground level	House	59° to 63°F
Reston, Virginia	Built into a hillside	Elementary school	Not available
University of Minnesota Williamson Hall	Below campus level	Bookstore	Never below 50°F
Kansas City, Missouri	Limestone caverns 50 to 200 feet deep	Industrial park	57°

6. What type of structure was built in limestone caverns?

7. In this table, the warmest natural temperature below ground level occurs in _____.

8. Which structure listed in the table would be called earth-sheltered?

    a. the school   b. the industrial park   c. the bookstore

PHYSICAL SCIENCE

Helicopters seem to ignore gravity. Lifted into the air by large spinning propellers, helicopters can do many things other aircraft cannot do. They can fly straight up or straight down. They can fly sideways, backward, and forward, and they can even hover or turn around in one spot. A helicopter can use almost any cleared space for takeoff and landing. This aircraft is especially valuable for rescue operations, aerial observation, and transportation in areas that have no roads or airfields.

WORDS TO KNOW

Biking with Science on Your Side
decrease, to become less
lubricated, made slippery or smooth
factor, a condition that brings about a result

Around the World in a Balloon
transparent, letting light rays through
propane, a flammable gas (widely used as fuel)

Does Your Car Have a Brain?
transistors, small, solid-state electronic devices
sensors, devices designed to detect, measure, and/or record information
analyzed, examined in detail

Magic Waves
weird, odd, strange
static, at rest, inactive
as a unit, as one
harness, to control

A Wonderful Tool: The Laser
detached, loose
delicate, needing careful handling
diffused, spread out, not concentrated
amplified, made stronger, increased power

Is There a Robot in Your Future?
advantage, superiority, favorable position
appendages, external parts

radioactive, giving off energy in particles or rays

A Steady and Welcome Wind
velocity, quickness of motion, speed
practical, usable, workable

Designing Cars Is not a Drag
resistance, a force or object that opposes motion
reduce, to make less
improve, to make better
model, a smaller representation of a full-sized object
vacuum, a space empty of a solid, liquid, or gas
retractable, able to be drawn back in

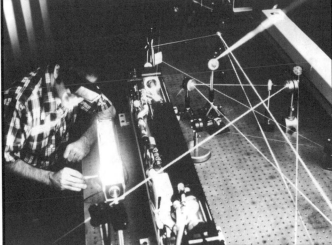

Biking with Science on Your Side

Be a champ! Win the race against friction.

The riders bend low over their bicycles and grip their handlebars. The signal is given for the 1.2-mile race to begin. Soon, the riders are whizzing along the track—first at 18 miles per hour, then at 30. They reach speeds of up to 40 miles per hour. Finally, speeding across the finish line at 50 miles per hour, comes the winner, Number Nine!

There is only one prize winner in a bike race, but all the riders win the race against friction (frĭk′shən). Friction is a problem that all moving objects have to overcome. It is a dragging force that opposes, or resists, the movement of one surface against another.

Friction is what makes sandpaper work. If you rub a piece of wood across a piece of sandpaper, some of the wood comes off. Friction is the force that wears away some of the wood. But if you rub the wood against a smooth piece of paper, no wood is worn away. The friction is less when two smooth surfaces move against each other. The amount of friction will also decrease when one or both objects are wet, or lubricated. Weight is also an important factor in the race against friction. The lighter the bike rider is, the less friction there is and the greater the speed.

A bike rider has to overcome air friction, or air resistance. When a bicycle racer bends over the handlebars, rider and bicycle make a smooth, low shape that can move faster through the air. Thin, hard bicycle tires also help reduce friction as the tires move over the track.

Of course, some friction is needed to get started and to brake, or stop. But the less friction, the more speed. And here comes Number Nine now!

1. Friction is a dragging _____ that opposes the movement of one surface against another.

2. A bicycle rider has to overcome _____ resistance.

3. The heavier a bike's rider is, the greater the amount of _____.

4. Which of the following would probably cause the most friction?
   a. rubbing two rough surfaces against each other
   b. rubbing a smooth surface against a rough one
   c. rubbing two smooth surfaces against each other

5. If a bike rider bends low over the handlebars, the amount of air friction she or he experiences
   a. increases.   b. decreases.   c. remains the same.

Use the table to answer questions 6, 7, and 8.

BRAKING DISTANCES

Distance a Car Will Travel after Driver Applies Brakes		
Car Speed (in miles per hour)	**Wet Surface** (in feet)	**Dry Surface** (in feet)
10	8	6
20	32	23.8
30	71.3	53.5
40	126.7	95
48	198	148.5
58	285	213.8
68	388	291

6. The greater the car speed, the _____ the car comes to a complete stop on either surface.
   a. more slowly   b. faster   c. more easily

7. Before coming to a complete stop, a car will travel _____ distance on a wet surface as compared with a dry surface.
   a. the same   b. a greater   c. a shorter

8. On a dry surface, at which speed below could a driver bring a car to a complete stop most quickly?
   a. 58 miles per hour   b. 40 miles per hour
   c. 20 miles per hour

Around the World in a Balloon!

For over two hundred years balloons have been used for travel. In 1999, the first around the world flight in a balloon was made. The adventurers took off from Switzerland, circled the Earth, and then landed in Egypt. How did they do it? Jacques Picard and Brian Jones traveled in combination helium and hot-air balloon called a "*Rozier*".

Balloons are made in different ways and use several means to ascend. *Ascend* means to rise upward. Hot-air balloons rise when the air inside is heated. This happens because the warmed air inside the balloon is lighter than the cooler air outside.

Hot-air balloons are also called aerostats. One type of aerostat has a transparent side. This allows the sun's rays to pass through it. The other side is made of black material. Because the transparent side is curved, it acts like a lens. This focuses the sunlight on the inner surface of the balloon. The air inside the balloon is then heated. The balloon has propellers on each side to rotate the balloon from one side to the other. Another type of hot-air balloon uses a propane burner to heat the inside air. You may have heard the rushing sound of the burner as a balloon passed overhead.

Some balloons use helium or hydrogen. Hydrogen and helium are lighter than air so they provide the force for ascending. A balloon filled with one of these gases will rise. How high and how much weight can be lifted depend on how big the balloon is. Both helium and hydrogen are used in weather balloons. These balloons are sent up daily to take weather measurements. Such measurements help forecast future weather.

1. Another word for a hot-air balloon is

      a.   aerostat.   b.   helium.   c.   hydrogen.

2. Sources of gas for a balloon include

      a.   oxygen and sodium.

      b.   wood and coal.

      c.   hydrogen and helium.

3. The curved, transparent side of the balloon acts like a _____.

4. Aerostats work on the principle that

      a.   hot air is heavier than cold air.

      b.   heated air rises.

      c.   cool air is lighter than warm air.

5. If you wanted to descend in a hot balloon, you would

      a.   quit heating the air inside.

      b.   provide more heat for the air inside.

      c.   find a cloud to cool the air inside.

6. When sunlight is absorbed on a black surface, it

      a.   cools off.   b.   rises.   c.   heats up.

Use the drawing to answer question 7.

7. You are the Aerostat's pilot. What must you do to go over the mountains?

Does Your Car Have a Brain?

A little wonder called the microprocessor is the "brain" behind the computer car.

The *microprocessor* (mī'krō prŏs'ĕs'ər) is a very small computer. It is a small, etched chip made of a chemical element called *silicon* (sĭl'ĭ kən). Chips are made in batches of thousands on wafer-thin sheets of silicon. Thousands of transistors and other electronic parts are etched onto these chips.

This is how the microprocessor works in a car. Located in a container behind the car's engine, the microprocessor is connected to sensors. The sensors monitor, or watch, the engine's cooling and exhaust systems as well as the flow of fuel. Information from the sensors will be fed into the microprocessor, where it is analyzed. If any part of these systems is not operating properly, that information is flashed onto a display screen mounted on the car's instrument panel.

The microprocessor is already being used for another purpose in some automobiles. The trip computer shows the driver information such as the car's current miles per gallon, the speed, the engine temperature, the estimated time of arrival, and the number of miles to go before the end of the trip. It even displays how many miles the car can go before running out of fuel.

Cars seem to get smarter all the time. Some cars have computers that "recognize" their drivers. The computer adjusts the seat to each driver's favorite position. The rear bumpers on other cars can "see." If the driver backs the car close to a wall or a person, a warning flashes on the instrument panel. The "brain" in a computer car is truly a little wonder.

1. The microprocessor is a very small _____.

2. The microprocessor is made of the chemical element _____.

3. Thousands of electronic parts are etched onto
   a. chips.
   b. wafers.
   c. boxes.

4. After information about the car's engine is analyzed, it may be
   a. monitored for correctness.
   b. fed into the microprocessor.
   c. flashed onto a display screen.

5. According to the story, the trip computer will not show the driver
   a. the best route to the destination.
   b. the number of miles per gallon of fuel.
   c. the approximate time of arrival.

6. This story would lead you to conclude that microprocessors are
   a. already in use in some cars.
   b. not yet in operation.
   c. a good idea but not very useful.

7. Which of the following uses of the microprocessor is most likely to reduce accidents?
   a. the engine-control system
   b. the rear bumper alarm
   c. the seat adjustment system

8. How would you rate the microprocessor's effect on the automobile industry?

1 2 3	4 5 6 7	8 9 10
NOT GREAT	GREAT	VERY GREAT

   a.   2   b.   6   c.   10

Magic Waves

Electromagnetic waves are not as weird as they might sound. Everyone has seen static electricity, such as when it makes hair stand on end. The magnetism of a magnet is static, too: it doesn't move by itself. It's only when electricity and magnetism move as a unit that they become a force called electromagnetism and can make waves.

There are other kinds of waves. The ocean has waves, of course. So does sound. And every kind of wave has the same basic form, something like a rounded "M." The high points of the wave are called the crests, and the space between them is called the wave's length, the wavelength.

But ocean waves travel on water, and sound waves travel on air. Electromagnetic waves can travel over or through anything, from empty space to solid rock.

Electromagnetic waves can be dangerous, since they create lightning and thunder. But we couldn't see without the electromagnetic waves that give us light. And when we harness electromagnetism, it can be very useful. We couldn't watch television, for instance, or cook with a microwave oven without electromagnetic waves.

How do we know about electromagnetic waves? A Scottish scientist named James Clark Maxwell first discovered them in 1865. A German scientist, Heinrich Rudolf Hertz, confirmed Maxwell's discovery in 1887, when he proved the existence of radio waves. Later scientists discovered that the amount of radiation given off by electromagnetic waves depends on the length of the wave. Harmless radio waves are the longest; light waves are shorter. The shortest electromagnetic waves, which take place within the nucleus—the heart—of the atom, produce the dangerous but useful radiation that allows doctors to take X rays.

1. _____ results when electricity and magnetism move together as a unit.

2. The distance between the crests of a wave is called the _____.

3. In paragraph 4 the word *harness* means

    a.   to make waves.

    b.   to control and use.

    c.   to create lightning and thunder.

4. Electromagnetic waves travel

    a.   on air.

    b.   on water.

    c.   over or through anything.

5. James Clark Maxwell was the first to discover

    a.   radio waves.

    b.   electromagnetic waves.

    c.   radiation.

6. The main idea of paragraph 4 is

    a.   electromagnetic waves are dangerous but useful.

    b.   electromagnetic waves are everywhere.

    c.   electromagnetic waves can't be seen.

7. According to the story, the shorter electromagnetic waves are, the

    a.   more useful they are.

    b.   less harmful they are to humans.

    c.   more harmful they are to humans.

A Wonderful Tool: The Laser

A doctor has discovered a detached *retina* (rĕt' n ə) in a patient's eye. In the past, a detached, or torn retina might have meant the loss of sight in the eye. But today, an amazing tool makes delicate eye surgery easier. The tool is the *laser* (lā' zər), a powerful beam of light. In certain cases, by carefully directing this light beam, the doctor can mend the retina.

Basically, the laser beam is the same kind of light that shines from a lamp bulb, but there are differences. The light from a bulb is diffused, or spread out, over a room. The light from a laser travels in a very narrow beam. The energy of the laser light is extremely intense and is focused in one direction.

Physicists have found that by shining light through certain crystals or gases, they can keep the light from spreading. The laser light becomes *amplified* (ăm'plə fīd'), or stronger, as two mirrors reflect it back and forth many times. Finally, the intensified light passes through a hole in the center of one of the mirrors. The word *laser* stands for "Light Amplification by Stimulated Emission of Radiation." You can see why it is easier simply to say "laser."

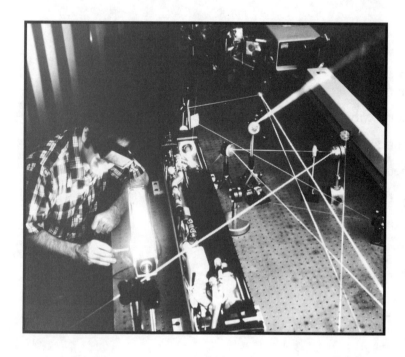

Today lasers are everywhere. At the supermarket checkout counter, laser technology finds the correct price and other information about your purchases. Every Compact Disc (CD) player has a small laser that it uses to read the information on the CD. In industry, a laser's mending and cutting ability is used to help make everything from electronic toys to heavy construction equipment.

Lasers are even used to send voice messages and television signals.

1. In the story, when something is amplified, it is made _____.

2. A Compact Disc (CD) player uses a small laser to _____ the information on the disc.

       a.   diffuse   b.   read   c.   mend

3. In this story, the doctor is going to use the laser beam to

       a.   sterilize instruments before an operation.

       b.   keep a patient from losing too much blood.

       c.   save a patient's eyesight.

4. One major difference between lamplight and laser light is that lamplight

       a.   is much brighter than a laser.

       b.   spreads out in all directions.

       c.   contains more heat than a laser.

5. Together, the mirrors in this article act like

       a.   a flashlight.

       b.   a surgeon's needle.

       c.   an amplifier.

6. How is laser light made stronger?

       a.   by passing it through a small hole

       b.   by reflecting it back and forth many times

       c.   by shining it through certain crystals or gases

Use the diagrams below to answer questions 7 and 8.

          A   **B**

7. Which diagram shows light energy in a very diffused form?

8. In which diagram would light energy be stronger?

Gliding Through Air

A colorful object floats over a hillside. Is it a bird? Is it a plane? No! It's a hang glider!

Recently, the sport of hang gliding has been growing in popularity. But the idea of building a glider that floats on air is actually very old. (A glider is a light, engineless aircraft with wide wings designed for long periods of soaring, or gliding.) For hundreds of years, people tried to find ways to fly like birds. Leonardo da Vinci designed a glider model in the fifteenth century. Over the years, many people died trying to fly gliders that were too large, too heavy, and too hard to handle.

Inventors tried many designs and made many improvements. Then, in the 1960s, the National Aeronautics and Space Administration (NASA) perfected a simple, lightweight wing design. The hang gliders you see today use NASA's design.

Hang gliders are now smaller and much easier to control than earlier models. The pilot hangs underneath the wings in a harness and holds on to a trapeze-like bar to steer the glider. If the pilot moves forward, the nose drops and the glider picks up speed. If the pilot moves backward, the nose rises and the glider loses speed. The pilot moves from side to side to turn the glider.

It sounds easy, but hang gliding can be very dangerous. Pilots must learn how to control their gliders in sudden updrafts of air. Sometimes hang gliders climb too steeply and the wings stall, or stop flying. A stall can lead to a crash. A crash may also occur if the glider makes too sharp a downward angle, causing the wings to collapse. But, with careful instruction, people can learn how to hang glide safely. Experienced pilots have traveled for up to 15 hours and as far as 29 miles.

1. According to the article, a glider is a light, _____ aircraft with wide _____ designed for long periods of gliding.

2. The idea of human flight

   a. is not very popular.

   b. has been around for hundreds of years.

   c. was proved to be impossible.

3. Why didn't early gliders work?

   a. The wings were much too small.

   b. They were hard to control.

   c. They were too light.

4. What part of the glider supports the pilot?

   a. the wings

   b. the harness

   c. the nose

5. When a person is hang gliding, what could be the result of either a sharp downward angle or a very steep climb?

Complete the following chart to answer questions 6, 7, and 8.

HANG GLIDING

Action Performed by Pilot	Result of Action
Moves forward	Nose drops, glider _____ speed. 6
Moves backward	Nose rises, glider _____ speed. 7
_____ 8	Glider turns.

Tuna Surprise

Hooking a bluefin tuna is like catching a fishing boat.

A tuna with the muscle strength of an adult man can swim for hours at 20 knots, or 20 nautical miles an hour. Scientists wanted to know how this is possible. They thought it might be the way a fish uses its tail. So they invented a robo-tuna to "swim" in a testing tank at the Massachusetts Institute of Technology.

Charlie Tuna, the robo-tuna, is controlled by overhead cables attached to his body. His tail moves back and forth in an S-pattern like a real tuna. After Charlie successfully swam the length of the tank, scientists began studying the water along his tail. They used tiny glow-in-the-dark bits, a laser, and a high-speed camera.

What the scientist saw surprised them. When Charlie swims, small *eddies*, or whirlpools form along his tail. At slow tail speeds, the eddies turn clockwise. At faster speeds, the eddies turn counter-clockwise. Clockwise eddies form a wake or drag. The counterclockwise eddies make *thrust* or jet. Tunas and other fish have been using jet energy for years!

Some day scientists hope to put this information to work. Submarine builders may be able to build a new self-propelled sub. It would be able to travel much longer than a battery-driven propeller sub. Its power would probably surprise Charlie.

1. Another word for *jet* is _____.

2. A tiny whirlpool is called an _____.

3. A bluefin tuna is as powerful as

    a. an adult man.

    b. a fishing boat.

    c. a jet plane.

4. In the list below, which thing did the scientists NOT use for their research?

    a. a camera

    b. a testing tank

    c. a robot tuna

    d. a laser

    e. a real tuna

5. A tuna's tail can make its own thrust

    a. at slow tail speeds.

    b. all the time.

    c. at higher tail speeds.

6. According to the story, the wake of a moving ship

    a. disappears at 20 knots.

    b. glows in the dark.

    c. slows it down.

7. Why are scientists studying Charlie Tuna?

    a. to learn about tuna

    b. to learn about swimming

    c. to learn how to build self-propelled submarines

Is There a Robot in Your Future?

Robots are the answer to the problem of how to fill many jobs in factories, hospitals, and police departments. Robots have certain advantages over human beings. They can do dangerous, difficult, or boring work day after day without getting tired.

Robots are machines that have flexible, or movable, *appendages* (ə pĕn' dĭj ĕz'). Appendages are parts that are attached to a main body. Many robots have arm-like appendages, making it possible for them to perform human tasks. Unlike most machines, robots can be "taught," or programmed, to do many difficult jobs in different ways.

Police departments use robots in place of human beings in bomb squads. Plants that handle radioactive materials also use robots because is also a threat to human safety. Car manufactures use robots to lift heavy parts and set them in place as cars move down the assembly line. All of these robots can produce high-quality work. Some of them may save human lives.

Hospitals use robots in operating rooms to assist doctors. The arm of this kind of robot is steadier than the hand of a human. The doctor can watch the work of the robot's arm on a

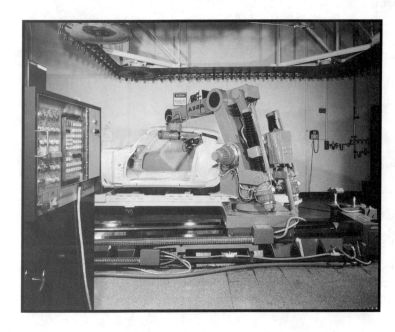

TV screen. Many of these operating room robots are controlled by the voice of the doctor, who still must tell the robot what to do.

Will robots ever be able to take over for human beings completely? Not really. After all, robots are nothing more than machines. People will still be needed to take care of, program, and supervise these machines. And, most important, robots cannot think. Thinking is a process that will still remain the property of human beings.

1. In this story, the word that refers to parts attached to a main body
    is _____.

2. In the story, the word *program* means the same as

    a.   produce.

    b.   instruct.

    c.   ask.

Decide whether the descriptions below refer to robots, humans, or both. Write the
letter R if the statement refers to robots, H if it refers to humans, and B if it refers
to both.

3. ____ They get bored with their jobs now and then.

4. ____ They never get tired from too much work.

5. ____ They can do more than one kind of job.

6. Why do you suppose doctors in the operating room needs to watch the
    robot's work on a TV screen?

7. An automobile factory uses robots to move heavy car parts from one place to
    another. One of the robots puts an entire set of parts in the wrong place. Who
    is responsible for this mistake?

    a.   the robot

    b.   the programmer

    c.   no one

8. An advertisement in the "Help Wanted" section of a newspaper states that the
    job demands a great deal of extra work, or overtime, and requires a lot of
    decision making. The manager would most likely hire

    a.   a human.

    b.   a robot.

    c.   either a human or a robot.

Moving Heat Around

A heat pump can be used to heat and cool buildings.

Imagine a machine that can *absorb* (ăb sôrb′), or soak up, heat and then pump the heat where it is needed. There is such a machine, and it is called a heat pump. Heating engineers think that the heat pump could be used to heat large buildings. This would mean that increasingly expensive fuels, such as gas and oil, could be put to other uses. A heat pump would be less expensive to use and could save energy, too.

The heat pump absorbs most of its heat from outdoors, even in winter. Everything, whether hot or cold, contains some heat. Cold things just have less heat than warm things. During winter, a heat pump absorbs heat from the air outdoors and transfers it inside to heat the air indoors. The heat pump must work *against* the principle that heat will always move from warm things to cold things. You can actually see this principle at work on a day. Look at the hood of a car that has been driven a long distance. The wavy lines you see over the hood are waves of heat leaving the hot engine and moving into the cooler air.

The heat pump works by reversing this natural flow of heat. It can do this because it is filled with a special liquid that absorbs heat. As it flows through tubes, the liquid absorbs heat from the outdoor air. Then the heated liquid boils and changes into a vapor or gas.

Next, the vapor is superheated and pumped through indoor tubes. Now that the vapor is hotter than room temperature, it condenses, turns back into a liquid. The pump removes the heat from the liquid and pumps it into the building's heating pipes.

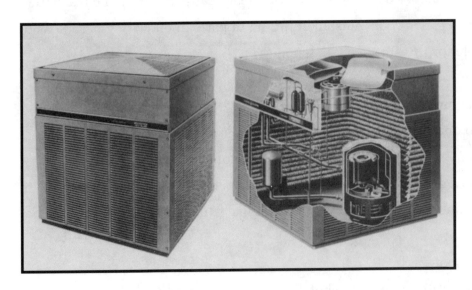

1. A heat pump is a machine that can soak up, or _____, heat from indoors or outdoors.

2. A heat pump gets most of its heat from _____.

3. According to the story, one important advantage of using a heat pump is that it can help save _____.

4. Unlike most other heating systems that use energy to *create* heat, a heat pump uses energy to _____ heat that is already available.

   a. save

   b. transfer

   c. store

5. What would happen if you changed a building's heat pump so that it worked in the opposite way?

   a. The building would get cooler.

   b. Nothing would happen.

   c. It would get colder outside.

6. In a glass of ice water, is the heat moving from the ice to the water or from the water to the ice? The heat moves from the _____ to the _____.

7. Which of the following would be an example of how heat flows naturally?

   a. snow melting

   b. a fan turning

   c. water boiling

8. A heat pump may be *twice* as efficient as a regular heating system when the temperature outdoors is 10°C, but only slightly more efficient when the temperature is 2°C. So areas where the winter temperature range is _____ would make the most efficient use of a heat pump.

   a. -10 to -2°C

   b. -2 to 2°C

   c. 2 to 7°C

Trash: Something of Value

There is treasure to be found in garbage dumps all across the country.

What treasure could be in rotting potato peels and moldy chicken bones? The treasure is called methane. Methane is an odorless, colorless gas that can be burned to obtain energy. This energy can be used to make electricity to light and heat buildings.

A garbage dump is a perfect factory for methane. The *organic* (ôr găn′ ĭk) waste materials found in garbage dumps serve as food for tiny bacteria. Organic matter comes from living things such as plants or animals. As the bacteria digest the organic garbage, they make methane. Over the years, as one load of garbage is piled on top of another, the amount of methane increases. Methane is a flammable gas, or one that is easily set on fire. In fact, some garbage dumps are so full of methane that they sometimes catch fire.

People in New Jersey found out about the treasure in their garbage dumps when they decided to turn one such refuse site into a large park. They knew they would have to get rid of the dangerous methane beneath the garbage. A gas company offered to buy the methane for its customers.

To get the methane out, the gas company drives long pipes deep into the garbage dump. Then the gas is pumped out into other pipes leading to people's homes. What some people sent out of their homes as garbage will go back into other people's homes as energy.

1. Matter that comes from living things such as plants or animals is called _____ matter.

2. In order for people to obtain energy from it, methane must be _____.

3. The energy obtained from methane can be used to make _____ to heat and light buildings.

4. The methane in a garbage dump _____ as the amount of garbage piles up.

       a.   increases   b.   decreases   c.   remains the same

5. It would be difficult to tell that a pipeline containing methane was leaking, because methane is _____ and _____.

6. Fill in the missing link in the chain of events below.

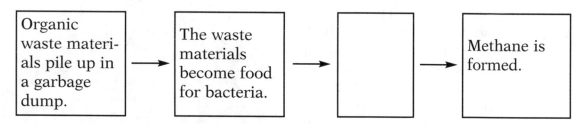

7. Which of the following would you be most likely to see on a sign at the entrance to a garbage dump?

       a.   Quiet Zone

       b.   No Smoking

       c.   Littering Prohibited

8. Methane would be classified as a kind of

       a.   organic waste.

       b.   gas.

       c.   bacterium.

The Wonder of Natural Gas

Natural gas has been used as a fuel to produce heat for thousands of years.

Natural Gas is a mixture of gases found deep in the earth. It is sometimes found in the same area as oil. The Chinese burned natural gas as a fuel three thousand years ago. They may have discovered the gas while drilling wells for salt or when some gas escaped from the ground and caught fire.

Finding natural gas in not easy. It involves the work of many people, including *geologists* (jē ŏl'ə jĭsts). Geologists are scientists who study the earth and its structure, or makeup. They use their knowledge of rocks and fossils to predict where natural gas may be found. Then wells are drilled to recover, or take, the gas from the ground. Recently, geologists helped to discover natural gas in the North Slope of Alaska.

Natural gas has also been found offshore. Some ocean gas wells are in water over 3 kilometers deep. Many gas wells have been drilled in the Gulf of Mexico, on the North Sea, and off the coast of South America. Once natural gas is recovered from a well, it is stored in underground tanks or transported through pipelines. Pumping stations along the pipeline help move gas along.

Natural gas can be used for many things, from cooking to air conditioning. Sometimes, it is burned as a fuel to create heat for generating steam. The steam then turns a turbine, which produces electricity.

1. Scientists who study the earth and its structure are called _____.

2. For thousands of years, people have used natural gas as a

    a.   mixture.

    b.   fuel.

    c.   fossil.

3. According to the story, natural gas is

    a.   easy to recover.

    b.   not found offshore.

    c.   difficult to find.

4. Natural gas is recovered by

    a.   drilling wells.

    b.   building pipelines.

    c.   studying rocks.

5. Burning natural gas produces

    a.   salt.

    b.   heat.

    c.   oil.

6. After natural gas is recovered, it is stored in underground _____ or transported through _____.

    a.   pumps, pipelines

    b.   pipelines, pumps

    c.   tanks, pipelines

7. In the chain of events below, what would the last event be?

| Natural gas is burned. | → | Heat is produced. | → | Steam is generated. | → | Steam turns a turbine. | → | |

Designing Cars Is Not a Drag

Can cars be designed to improve gas mileage.

Automobile designers are always looking for ways to get good gas mileage in cars. One way to do this is to make the cars more streamlined.

Automobile designers know that wind resistance slows down objects when they move through the air. The force that slows down an object moving through the air is called drag. As an example, a parachute has a lot of drag. A racing car has less drag than a parachute.

Much of a car's fuel is used to overcome drag. Studies show that at 50 miles an hour, 50 percent of a car's fuel and horsepower is used to overcome drag. If designers can reduce drag they can improve on the car's gas mileage.

Designers use wind tunnels to study the drag conditions of automobiles. The wind tunnel has a width of between 35 to 75 feet and is as long as 800 feet. The object tested can be a full-size car or a model car.

Inside the tunnel is a large fan and a test car. The fan blows air directly over the car at about 50 miles an hour. During the test, the designers observe how the air blows over the windshield, headlights, side trim, roof, hood, and the underside part of the car.

The wind tunnel studies have shown several trouble spots. The underside part of the car produces about 20 percent of the drag in most cars. The rear of the car is also a problem. The air that flows from the front of the car collects in the back creating a vacuum which pulls the car backward. Even windshields, side mirrors, mud flaps, and license plates cause drag. From the studies, designers have made improvements. They made car hoods slope down to act as a wedge against the wind. Other improvements include small radiator grills and retractable headlights.

1. Much of a car's fuel is used to overcome _____.

2. In the first paragraph, *"streamline"* means
    a. less resistant to moving air and water.
    b. more resistant to moving air and water.
    c. very strong while it moves through air and water.

3. About _____ percent of a car's fuel and horsepower is used to overcome drag.
    a. 30
    b. 50
    c. 70

4. The wind tunnel is used to study the
    a. drag conditions of an automobile.
    b. engine performance of an automobile.
    c. gas mileage of an automobile.

5. According to the story what part of the car produces 20 percent of the drag?
    a. the hood of the car
    b. the underside part of the car
    c. the roof of the car

6. From what you read in the story, which object of the same size would probably have the most drag?
    a. a round cylinder
    b. a square block
    c. a cone-shaped object

7. The main purpose of reducing drag in automobiles is to
    a. improve gas mileage of automobiles.
    b. make the automobile look more modern.
    c. make the automobile stronger.

8. If you were building a hang glider, you would probably want
    a. a lot of drag.
    b. no drag at all.
    c. very little drag.

ENVIRONMENTAL SCIENCE

There are more than 60 kinds of deer. In North America the best known are moose, elk, caribou, mule deer, and white-tailed deer. There are now more white-tailed deer than when the Pilgrims landed. Humans have greatly contributed to the great increase in white-tail deer. With restricted hunting, the disappearance of wolves, bears, and coyotes, and the growth of the suburbs with lush bushes, trees, and grasses, the deer population has exploded. In many places they are destroying plants and are hazards to traffic. The white-tail deer is an example of the challenge humans face in helping restore a balance in nature.

WORDS TO KNOW

China's Great Dam

reservoir, a body of water collected and stored in a natural or artificial lake

disastrous, widespread destruction, ruinous

hydroelectric power, electricity produced by energy of flowing water

aquatic, living or growing in or on the water

habitats, the environment where an organism normally lives

What's Happening to the Ozone Layer?

oxygen, an odorless, tasteless element that makes up about 21 percent of the Earth's atmosphere

atom, the smallest unit of an element such as oxygen

radiation, the emission of particles or waves from light or heat

Save the Wolf!

endangered, in danger of no longer existing

extinct, no longer alive

predators, animals that live by eating other animals

Constant Energy from the Sun

solar, relating to the sun

conversion, the changing of something into another form

installed, set in position

manufacturing, to make finished products from raw materials

Harnessing Steam from Volcanoes

harness, to bring under control and direct the force of

geothermal, pertaining to the internal heat of the Earth

imported, brought from outside the country

alternative, allowing a choice of two or more things

A Steady and Welcome Wind

turbine, machine spun by flowing water or wind to create electricity

peninsula, land with water on three sides

velocity, speed

clusters, groups

Clean Water Makes a Difference

tidewater, water along shores and in some streams along coastlines that are affected by tides

estuary, an inlet or arm of the sea; also, the wide mouth of a river where the tide meets the current

nutrients, the parts of foods used by plants and animals for growth and energy

eliminate, to get rid of

sensitive, easily hurt

Alternative Fuels

efficient, acting or causing an effect with little waste

Getting Oil from Sand

deposits, something put down in a
  layer, or layers, by a natural process
conveyer, a machine, such as a
  moving belt, that moves bulk
  materials from one place to another

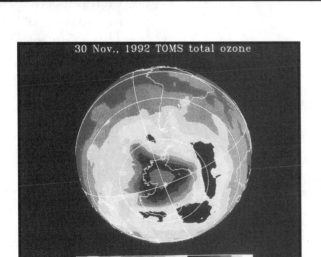

30 Nov., 1992 TOMS total ozone

Total DU

China's Great Dam

A new dam is being built in China. It is called the Three Gorges Dam. When completed in 2009, the dam will be about 610 feet high. The new dam will be so large that it will be seen from space.

The Three Gorges Dam is being built on China's Yangtze River. The Yangtze River is the third longest river in the world. It stretches almost 4,000 miles long. Building the dam across the Yangtze River will create a large lake, called a reservoir. The reservoir will be approximately 375 miles long and up to 600 feet deep. To create a reservoir of this size and depth, about 1,000 villages along the river will be flooded. Between one and two million people will need to be resettled from their river towns and villages.

The Yangtze River has caused disastrous floods throughout the history of China. Many lives have been lost. Villages have been destroyed. The new dam will control flooding. It will store excess water during times of heavy rainfall and runoff. The dam will release water during drier periods. Large deepwater ships will be able to navigate on the river, too. These vessels will be able to travel 1,500 miles into the interior of China. The Three Gorges Dam will also provide hydroelectric power for electricity.

Some environmentalists are not in favor of the Three Gorges Dam. They point out that the construction of the dam will destroy many aquatic habitats. The dam's reservoir may become polluted. The pollution could contaminate drinking water supplies and cause outbreaks of disease.

QUESTIONS

1. The Three Gorges Dam in China will provide flood control, a navigation route for boats, and _____ for electricity.
2. The word in the article that means "a structure that restricts the flow of water" is a _____.
3. The Yangtze River is the _____ longest river in the world.
    a. second   b. third   c. fourth
4. The Three Gorges Dam will be completed by the year 2012.
    a. True   b. False   c. Article does not say.
5. One of the environmental problems with the building of the dam is that
    a. people will be resettled from their villages.
    b. aquatic habitats will be destroyed.
    c. boats will not be able to navigate on the river.

Use the map to answer questions 6, 7, and 8.

6. The Yangtze River flows into the
    a. Pacific Ocean.   b. South China Sea.   c. Yellow Sea.
7. The Three Gorges Dam is between
    a. Shanghai and Yichang.   b. Yichang and Chongqing.
        c. Hong Kong and Chongqing.
8. A country located on the southern border of China is
    a. Cambodia.   b. Vietnam.   c. Thailand.

What's Happening to the Ozone Layer?

The Earth's ozone layer is important to our health.

Scientists have discovered holes in the Earth's ozone layer. The ozone layer is located in the Earth's stratosphere. The stratosphere is a layer of the atmosphere. The stratosphere is between 6 and 31 miles above the Earth's surface. Large passenger airplanes fly in the stratosphere to avoid bad weather and strong winds.

Ozone is an oxygen molecule. It is composed of three oxygen atoms. The oxygen we breathe is composed of two oxygen atoms. About 90 percent of all ozone is located in the stratosphere.

The ozone layer acts like a blanket that protects plants and animals from the sun's radiation. Too much radiation can cause skin cancers in humans and can harm other living organisms. For several years, the ozone layer in the stratosphere has been thinning out. More radiation is entering the Earth's surface, and scientists are very concerned.

One of the largest holes is above Antarctica. The hole in the ozone layer occurs yearly in late September. The hole is roughly the size of the continental United States. Ozone holes have also appeared in the Arctic and in parts of Russia.

The thinning out of the ozone layer, is caused by human-made chemicals. Some of these chemicals enter the stratosphere and destroy parts of the ozone layer. The United States and other countries have agreed to put a ban on using chemicals that destroy the ozone layer. Scientists hope these chemicals will disappear in the future.

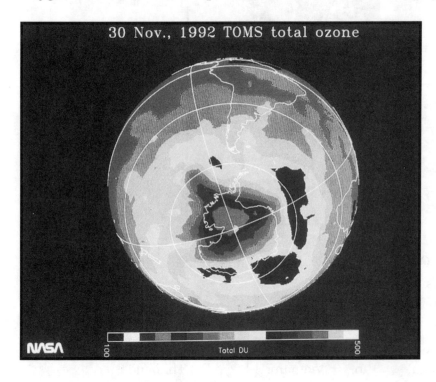

30 Nov., 1992 TOMS total ozone

NASA

Total DU   100   500

1. The word in the story that means "a group of atoms held together" is a _____.

2. Ozone contains _____ atoms of oxygen.
   a.   three   b.   four   c.   two

3. The ozone layer protects plants and animals from
   a.   radiation.   b.   sound waves.   c.   radio waves.

4. The ozone holes have only appeared in Antarctica.
   a.   True   b.   False   c.   The story does not tell

5. Most ozone is found
   a.   on Earth's surface.   b.   in the stratosphere.
   c.   in Antarctica.

6. Too much radiation from the sun can cause
   a.   dizziness in humans.   b.   heart problems in humans.
   c.   skin cancer in humans.

Use the map to answer questions 7 and 8.

GRAY AREA REPRESENTS OZONE HOLE

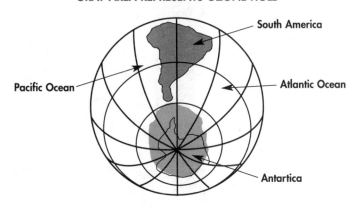

7. Antarctica is closest to
   a.   South America.   b.   North America.   c.   Asia.

8. The ozone hole in the stratosphere covers
   a.   none of Antarctica.   b.   some of Antarctica.
   c.   much of Antarctica.

Save the Wolf!

How do wolves keep other animals from starving?

Many years ago, large, dog-like mammals called wolves lived and hunted in many parts of Mexico, Canada, and the United States. But by the early 1970s , wolves had disappeared from many places. The red wolf, for example, was greatly reduced in numbers. Many of these wolves were hunted and killed. Also, no steps were taken to keep the wolves from breeding with their western relatives, the coyotes. Today, wolves are an endangered species in the lower United States except in Minnesota. The wolves are in danger of becoming *extinct*, or dying out completely.

Scientists feel that wolves should be protected because they are an important link in a *food chain*. Every living plant or animal is part of a food chain. A food chain begins with the sun. Green plants use sunlight to make their own food. Then the green plants are eaten by small animals, such as rabbits, which are then eaten by larger animals.

Grass, deer, and wolves are an example of one food chain. Wolves are an important part of the chain because they are *predators*. Predators get their food by hunting other animals. Wolves hunt animals such as deer, elk, and caribou. By doing this, they help preserve the balance of nature.

If the wolf is not saved from extinction, elk and deer herds may increase to such an extent that, in time, these animals might eat up all the available plant food in an area. As a result, many animals would starve to death.

Wolves keep the population of a herd in balance. They control the size of the herd by hunting and feeding on the old, crippled, and sick animals. So predators such as wolves play an essential part in a food chain and help to preserve wildlife.

In the 1990s, the U.S. Fish and Wildlife Service reintroduced gray wolves to Yellowstone National Park. Their plan was to reestablish the species in the park after an absence of 60 years.

1. An animal that is in danger of dying out completely could soon become _____.

2. Animals that get their food by hunting other animals are called _____.

3. Scientists are concerned with protecting the wolf because it helps _____ the balance of nature.

4. In the food chain described in the article, wolves play an important role in the _____ of the food chain.

    a.   first chain   b.   second stage   c.   third stage

5. Which of the following has not led to a decrease in the number of red wolves?

    a.   letting them breed with coyotes

    b.   poisoning them with plant spray

    c.   hunting and killing them

6. Which of the following animals would most likely be a predator?

    a.   a deer   b.   a fox   c.   a rabbit

Use the table to answer questions 7 and 8.

Group	Endangered U.S.	Endangered Foreign	Threatened U.S.	Threatened Foreign	Total Species
Mammals	61	251	8	16	336
Birds	75	178	15	6	274
Reptiles	14	65	21	14	114
Amphibians	9	8	8	1	26
Fish	69	11	41	0	121
Snails	18	1	10	0	29
Insects	28	4	9	0	41
Animal Subtotal	274	518	112	37	941

7. On this list, which group is in *least* danger of becoming extinct?

8. There are more kinds of endangered _____ in the United States than in foreign countries.

    a.   birds   b.   snails   c.   mammals

Green Sea Turtles Make a Comeback

Green sea turtles are reappearing in ocean waters throughout the world.

Sandy beaches are the breeding grounds of the green sea turtle. Breeding grounds are places where animals and fish go to mate and lay their eggs. The green sea turtle digs a nest in the sandy beach away from the water's edge. After laying about 100 eggs, the turtle covers the nest with sand. About two months later, the eggs hatch and the baby turtles, each about 3 inches long, immediately head for the sea. On their way, the newly hatched turtles are preyed upon by hungry birds, crabs, or raccoons. Even those sea turtles that reach water are not always safe. Some of them are eaten by hungry fish along the shore.

Green turtles are found in tropical or warm regions throughout the world. One place where green turtles lay their eggs is on islands in the Sulu Sea off the coast of Malaysia. For years, island people had collected turtle eggs because their shells fetched a good price and they could be used in cooking. In fact, so many eggs were collected that fewer and fewer turtles hatched each year, and the green turtle was in danger of becoming extinct (ĭk stĭngkt'), or disappearing completely from the Earth.

So the government of Malaysia decided to buy the islands and name them Turtle Islands National Park. The eggs were now protected by wildlife services and game wardens. Records were kept of the number of eggs laid and the turtles that hatched. During a six year period, more than one million green sea turtles made it to the water!

1. Places where animals and fish go to mate and lay their eggs are called _____.

2. Baby sea turtles are about _____ inches long.

3. Newly hatched green sea turtles are in danger of being

    a. buried in sand.

    b. preyed upon.

    c. drowned.

4. The green sea turtle seems to prefer a _____ climate.

5. What led to the reduction in the number of green sea turtles hatched on the islands in the Sulu Sea?

    a. People collected the turtle eggs to sell.

    b. Turtle eggs were washed away by ocean currents.

    c. Fish came ashore and ate the turtle eggs.

6. The Malaysian government bought the islands in the Sulu Sea because it wanted to save the wildlife from extinction.

    a. True

    b. False

    c. The article does not say.

7. Of the following places, you are most likely to find green sea turtles off the coast of

    a. Maine.   b. Oregon.   c. Florida.

8. How would you rate the chances of the Turtle Islands' baby sea turtles for reaching the sea?

1 2 3	4 5 6 7	8 9 10
FAIR	**GOOD**	**EXCELLENT**

    a.   3   b.   6   c.   9

Constant Energy from the Sun

As electric bills increase and fossil fuels decrease, free and plentiful sunlight looks brighter and brighter.

Radiation from the sun can be converted, or changed, directly into electricity in several ways. But, at the moment, the most promising way makes use of a device called the photovoltaic, or solar, cell. Developed in the early 1950s, photovoltaic cells have been widely used aboard space vehicles and as battery chargers aboard boats. Photovoltaic cells are made from silicon crystals that convert sunlight to electricity. Simply stated, the conversion of solar radiation into electrical energy takes place when sunlight hits certain materials in the photovoltaic cell and forms an electric current.

Some home builders in California used special shingles made from photovoltaic cells. The shingles were installed as part of a watertight roof. The solar cell shingles are a pollution-free source of energy.

The photovoltaic roof shingles were mounted on portions of the south-facing roof of each home. The photovoltaic roof shingles generated about two kilowatts, or about 60 percent of each home's total electric needs. The balance of the home's need for electricity was supplied from a local utility company.

Homes with photovoltaic roofs can also generate power to recharge electric cars. Electric cars are a totally emission-free source of transportation.

In recent years, improved designs of solar cells have lowered their manufacturing cost. However, solar cells still provide electricity at a higher cost than power from utility companies. But solar experts agree that homeowners could save money in the long run with their own rooftop power plants.

1. Another name for *photovoltaic* cell is _____.

2. The purpose of the photovoltaic cell is to convert solar radiation to _____.

3. All solar-energy systems are dependent upon
    a. the sun.
    b. electrical energy.
    c. photovoltaic cells.

4. The photovoltaic roof shingles generate about _____ percent of a house's electric needs.
    a. 20   b. 40   c. 60

5. Compared to the cost of electricity supplied from a utility company, photovoltaic electricity
    a. costs the same.   b. costs less.   c. costs more.

6. The photovoltaic roof shingles can also generate power to recharge
    a. power boats.   b. electric cars.   c. motorcycles.

7. In order to generate most of a home's electric needs, photovoltaic roof shingles would need to generate about
    a. 10 more kilowatts.
    b. 1 more kilowatt.
    c. 100 more kilowatts.

8. Fill in the missing link in the chain of events below:

    | Photovoltaic Cells | → | | → | Electrical Energy |

    a. Microwaves   b. Solar Panels   c. Sunlight

Harnessing Steam from Volcanoes

Can heat energy deep inside Earth provide electrical energy for a city?

The Hawaiian Islands are actually the tops of volcanoes that stick up above the surface of the Pacific Ocean. From time to time, one of the active volcanoes may erupt, and lives and property may be endangered. However, Hawaiians are trying to make good use of their volcanoes.

On the largest island, Hawaii, steps are being taken to harness some of the *geothermal* (jē'ō thûr'məl) energy found in volcanoes. The word *geothermal* refers to any form of heat energy that comes from deep inside the Earth. This heat may be in the form of hot water, hot rocks, or steam. One hot well was found at a depth of about 6,270 feet, and the steam pouring out of it had a temperature of 670° F!

Imported oil is used to supply 90 percent of Hawaii's energy needs. Hawaiians are now considering alternative energy sources to reduce their dependence on oil. One of the alternative energy sources is geothermal energy. Steam from hot wells can be used to drive the generators in the plants, and the generators will produce electrical energy. Geothermal energy does not produce air pollutants associated with the burning of oil. Hawaii's other alternative energy possibilities include wind and solar energy.

QUESTIONS

1. The word used to describe any form of heat energy that comes from deep inside the Earth is _____.

2. Hawaiians are planning to use the energy from _____ to produce electricity.

3. The energy to be harnessed from geothermal energy can be in the form of
    a.   oil.   b.   hot water.   c.   air.

4. What type of plant produces electricity?
    a.   a geothermal plant
    b.   an oil plant
    c.   a generating plant

5. The article suggests that when engineers are drilling wells, they
    a.   drill very deep into the Earth.
    b.   use generators to drill through rocks.
    c.   measure the steam from the well.

6. Which of the following is classified as a form of geothermal energy?
    a.   wind   b.   steam   c.   oil

Use the map below to answer question 7.

Geothermal Resources—Areas of Promise

• HOT SPRINGS

7. Geothermal energy can be obtained from hot springs as well as volcanoes. What part of the continental United States shows the greatest promise as a source of geothermal energy?
    a.   the West   b.   the East   c.   the North

A Steady Welcome Wind

Denmark was known as a land of windmills for many years. But, by the 1930s, most of them were no longer used because it seemed old-fashioned. Yet the winds from the North Sea continued to blow across Denmark's flat Jutland peninsula at a high *velocity* (və lŏs'ĭtē), or speed.

Today, this steady north wind is welcome. With a velocity of over 3 1/2 miles per hour, it makes wind-generated electricity practical. Scientists estimate that the wind blows about 300 days a year. And since there are few forests and no mountains in Denmark to break the wind, it is a powerful wind, indeed.

On the western edge of Jutland, a giant windmill is now at work. The windmill was built by citizens in the area who wanted to prove to the world how wind can be converted, or changed, into energy cheaply. The windmill stands 265 feet high in Ulfborg, Jutland. Sitting atop a 54-meter-high tower are three giant fiberglass propellers, each 90 feet long. The tower is made of concrete and steel at the top. The fiberglass blades can bend but will not break in strong winds which can reach hurricane strength.

Electrical power is measured in a unit called the *kilowatt* (kĭl'əwŏt'). A kilowatt-hour is the use of 1 kilowatt of power for one hour. The Jutland windmill can produce 4 million kilowatt-hours of electrical power in one year. That is enough to provide electricity for 2,800 people. There are also several smaller windmills around the country. At Windmill Park near the city of Ebeltoft, a cluster, or group, of windmills provides electricity for many nearby families.

1. In the story, the word that means "speed" is _____.

2. Electrical power is measured in a unit called the _____.

3. Denmark's wind comes from the _____.

4. According to the story, Denmark is a good place to use wind power because there are

   a. many windmills left over from the 1930s.

   b. no mountains and few forests to slow down the wind.

   c. fewer people to provide with electrical power.

5. What makes fiberglass a good material for the propeller blades on a windmill?

   a. It is flexible but strong.   b. It is inexpensive to use.

   c. It allows light to pass through.

6. Under which of the following headings would you list the Jutland windmill?

   a. Citizens' Efforts

   b. Free Electrical Power

   c. An Unsteady Tower

Use the table to answer questions 7 and 8.

AVERAGE WIND SPEEDS AT SOME WEATHER STATIONS IN THE UNITED STATES (computed through 1977)

Location	Average Speed (in miles per hour)	Location	Average Speed (in miles per hour)
Bismarck, N.Dak.	10.1	Detroit, Mich.	9.8
		Galveston, Tex.	10.6
Boston, Mass.	12.0	Key West, Fla.	17.9
Buffalo, N.Y.	11.8	Minneapolis,	
Cape Hatteras, N.C.	11.2	Minn.	10.1
		Omaha, Nebr.	11.3
Chicago, Ill.	10.0	San Francisco,	
Cleveland, Ohio	10.4	Calif.	10.1

7. In which location would a wind-powered generator be least efficient?

8. In which two cities is the velocity of the wind the greatest?

Clean Water Makes a Difference

Polluted water: People cannot drink it. Fish cannot live in it. Plants cannot "breath" in it.

In the late 1960s, the life in rivers, lakes, and tidewater areas was dying. Chemicals, raw sewage, and pesticides were dumped into the water. Both visible and invisible pollutants filled the rivers, and there was serious concern that pollution would eventually ruin many bodies of water all over the world.

Pollution is especially damaging to a body of water called an *estuary* (ĕs′choo ĕr′ē). An estuary is a bay at the mouth of a major river where salt-water and fresh water mix. Estuaries are rich in nutrients and support the growth of great amounts of *phytoplankton* (fī tō plăngk′ tən). Phytoplankton—tiny, floating plants—serve as a source of food for young fish. Many large cities, such as New York, Boston, San Francisco, and London (England), are located on estuaries. Household and industrial wastes from cities are sometimes dumped into the estuaries. Small amounts of waste fertilize the bays, but large amounts can actually poison the water life.

Today, the results of laws to eliminate pollution can be seen in many places. For 162 years the Kennebeck River in Maine was filled with pollutants. But since June 2000, salmon, which are especially sensitive to pollution, have been caught in the river. Lake Erie also shows signs of life. A rugged fish called the alewife has returned to the waters of Lake Erie. In time, perhaps all of the world's waters will be rescued.

1. A body of water at the mouth of a major river where salt water and fresh water mix is called an _____.

2. In the article, the tiny, floating plants that are a source of food for young fish are called _____.

3. Three forms of pollution mentioned in the story are chemicals, raw sewage, and

    a. fertilizers.   b. pesticides.   c. nutrients.

4. According to the story, we know that the pollution in the Kennebeck River has cleared up because

    a. some salmon have been caught in the river.

    b. there is more plant growth in the river.

    c. the water *looks* much cleaner.

5. Dumping very small amounts of waste into an estuary

    a. helps encourage the growth of phytoplankton.

    b. poisons the water forever.

    c. kills the source of food for young fish.

6. One sure sign that the pollution in a body of water is decreasing is when

    a. you can no longer see any pollutants in the water.

    b. plants cannot float on the water.

    c. the plant and animal life return to the water.

7. You could conclude from the story that water pollution is

    a. a minor problem.

    b. an international problem.

    c. no longer a problem.

8. Fill in the missing link in the chain of events below.

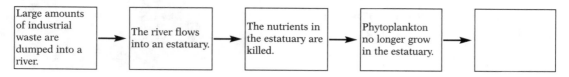

Alternative Fuels

Methanol and liquid hydrogen may be fuels of the future.

Methanol is a fuel produced by mixing about 10 percent alcohol with 90 percent gasoline, which is a product of oil. The mixture is used to power automobiles. Unlike the oil used to produce gasoline, alcohol is a renewable resource made from plants. Some sources of alcohol are sugar, wheat, potatoes, and corn.

By using methanol, the United States can reduce the amount of oil it imports from other countries. Experts estimate that we may save as much as 200 million barrels of oil each year. Methanol has another advantage over gasoline. Because methanol burns cleaner than gasoline, it produces fewer pollutants. Also, in engines that are designed to use it, methanol is about as efficient as gasoline.

A second new fuel source is hydrogen. Unlike oil, hydrogen is plentiful. Also, very little pollution is created when hydrogen burns. In fact, when hydrogen mixes with oxygen during burning, ordinary water is formed. All water contains some hydrogen, so hydrogen can be obtained from many different sources, even seawater.

One way to separate hydrogen from water is by *electrolysis* (ĭ lĕk trŏl'ĭ sĭs). Electrolysis is a process in which a current of electricity is passed through water, causing the water molecules to split into hydrogen and oxygen. The hydrogen can then be collected, changed into liquid form, and transported easily and cheaply by pipeline or tanker. Pipelines could bring hydrogen directly to homes and factories for heating purposes and to "gas" stations for use in cars.

Use the list of words below to complete questions 1 through 4.

methanol   fuel   electrolysis   alcohol

1.   A current of electricity is passed through water, separating it into hydrogen and oxygen. This process is called _____.

2.   In the story, a mixture that burns cleaner than gasoline is _____.

3.   Something that is burned to produce energy is called a _____.

4.   In the story, a renewable resource made from plants is _____.

5.   How will the use of methanol affect oil imports to the United States?

      a.   Oil imports will increase.

      b.   Oil imports will decrease.

      c.   Oil imports will remain the same.

6.   According to the story, when compared with gasoline, methanol

      a.   is about as efficient.

      b.   costs much more.

      c.   produces pollutants.

7.   The hydrogen fuel discussed in the story would come under the heading of

      a.   Liquids.

      b.   Solids.

      c.   Gases.

8.   Hydrogen gas is lighter than gasoline, but it takes up a lot of space. In what way would this be a problem with today's cars?

      a.   The cars' engines would need to be heavier.

      b.   The cars' fuel tanks would have to be enlarged.

      c.   The cars would need larger pollution-control devices.

Getting Oil from Sand

Have you ever heard of mining oil?

A black, sticky, tar-like substance is now being mined in Alberta, a province in western Canada. This substance, called bitumen (bĭ tōō′mən), is found in tar-sand deposits near the Athabasca River. It is mined in much the same way that coal is obtained from the ground through a process known as strip mining.

The tar-sand deposits contain a heavy oil that binds the sand together. The deposits lie beneath a layer of decaying vegetation. This plant matter is scraped away by giant bucket-shaped shovels. Then the shovels scoop up the tars and deposits. Next, the sticky tar-sands are dumped onto conveyer belts that feed the sand into large tanks. In these tanks, hot water, steam, and air separate the bitumen from the sand. The bitumen is then chemically treated so that only a crude oil is left. To produce 1 barrel of crude oil, about 2.2 tons of the tar-sand material must be processed.

It is very costly to produce crude oil from the bitumen. However, with the rising costs of imported oil and the alarming rate at which it is being used, mined crude oil may well become an important energy source in the near future.

In the United States, Utah and California also contain deposits of tar sands. But it is believed that even more of this material is required to produce a single barrel of crude oil. So attention is still focused on the huge, rich Alberta deposits, sometimes referred to as the Saudi Arabia of tar sands.

1. The black, sticky, tar-like substance needed to make crude oil is called _____.

2. The Canadian province where tar-sand deposits are found is _____.

3. What binds the sand together?

4. What must be removed first before miners are able to get the tar-sand deposits?

   a. decaying plants   b. chemicals   c. steam

5. After being separated from the sand, the bitumen gets

   a. dumped onto a conveyer belt.

   b. placed in huge tanks.

   c. chemically processed into crude oil.

Use the graphs below to answer questions 6, 7, and 8. The graphs show the kind of energy produced by the United States in 1977 and 1987.

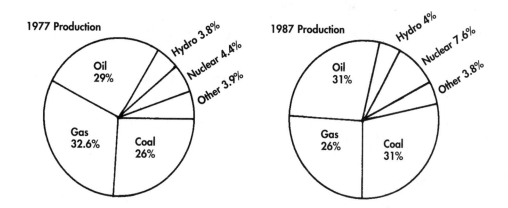

6. What was the largest kind of energy production in 1977?

   a. coal   b. gas power   c. water   d. oil

7. Which kind of energy production has grown the most since 1977?

8. By how much has natural gas production changed since 1977?

BIBLIOGRAPHY

Books on Life Science

_____. *Human Body*. New York: Time-Life, 1992.

Bright, Michael. *Pollution & Wildlife, Survival Series*. New York: Gloucester Press, 1987.

Burton, Robert. *Beginnings of Life*. Maryknoll, NY: Orbis, 1986.

Cochrane, Jennifer. *Land Energy, Project Ecology Series*. New York: Bookwright Press, 1987.

Henderson, Douglass. *Dinosaur Tree*. New York: Simon & Schuster, 1994.

Lorimer, Lawrence T.; Fowler, Keith. *The Human Body: A Fascinating See-Through View of How Bodies Work*, Pleasantville, N.Y.: Readers Digest, 1999.

McLaughlin, Molly. *Earthworms, Dirt, & Rotten Leaves: An Exploration in Ecology*, New York: Atheneum, 1986.

Parker, Steve. *Charles Darwin and Evolution*. New York: Harper-Collins, 1992.

Rivers, Lynn; McDonald, Sharon. *Pigs, Plants & Other Biological Wonders*. Dubuque, Ia: Kendall/Hunt Publishing, 1999.

Selsam, Millicent E. and Joyce Hunt. *A First Look at Owls, Eagles, & Other Hunters of the Sky, A First Look At...Series*. New York: Walker, 1986.

Silverstein, Alvin; Silverstein, Virginia; Silverstein, Laura. *Common Colds (My Health)*. New York: F. Watts, 1999.

Stockley, Corinne. *Animal Behavior, Science and Nature Series*. Tulsa, OK: EDC Publications, 1992.

Waldbauer, Gilbert. *Millions of Monarchs, Bunches of Beetles: How Bugs Find Strength in Numbers*. Cambridge, Mass.: Harvard Univ. Press, 2000

Books on Earth-Space Science

_____. *Planet Earth*. New York: Time-Life, 1992.

_____. *Space Planets*. New York: Time-Life, 1992.

Branley, Franklin M. *Shooting Stars*. New York: Crowell, 1989.

_____. Branley, Franklin M. *Sunshine Makes the Seasons*, rev. ed. New York: Crowell, 1986.

_____. Branley, Franklin M. *The Sun, Our Nearest Star*. New York: Crowell, 1988.

Fisher, Leonard Everett. *Galileo*. New York: Macmillan, Maxwell Macmillan Canada; Maxwell Macmillan International, 1992.

Kahl, Jonathan D. W. *Storm Warning: Tornadoes and Hurricanes* (How's The Weather?). Minneapolis, Minn.: Learner Publications, 1993.

Lampton, Christopher F. *Volcano*. Connecticut: Millbrook Press, 1991.

Pollard, Michael. *Air, Water, Weather*. New York: Facts on File, 1987.

Massa, Renato. *Along the Coasts* (Deep Blue Planet). Texas: Raintree-Steck-Vaughn, 1997.

Segal, Justin; Lyon, Carol. *The Amazing Space Almanac*. Illinois: NTC Publishing Group, 1999.

Scott, Elaine. *Adventure in Space: The Flight to Fix the Hubbel*. New York: Disney Press, 1998.

Web-sites on Earth-Space Science

Venus/NASA/Magellan Mission to Venus
http://nssdc.gsfc.nasa.gov/planetary/magellan.html

Pioneer Venus Orbiter
http://www.nssdc.gsfc.nasa.gov/planetery/pioneer-venus.html

Books on Physical Science

_____. *Physical Forces*. New York: Time-Life, 1992.

Berger, Melvin. *Atoms, Molecules, & Quarks*. New York: Putnam, 1986.

Cwiklik, Robert. *Albert Einstein and the Theory of Relativity* (Barrons Solution Series). New York: Barrons Juveniles, 1987.

Glover, David; Lloyd, Frances. *The Super Science Book of Sound*. Texas: Raintree-Steck-Vaughn, 1994.

Morgan, Nina. *Lasers* (20th Century Inventions). Texas: Raintree-Steck-Vaughn, 1997.

Books on Environmental Science

Bright, Michael. *Pollution & Wildlife, Survival Series*. New York: Gloucester Press, 1987.

BIBLIOGRAPHY

Cochrane, Jennifer. *Land Energy, Project Ecology Series*. New York: Bookwright Press, 1987.

Forman, Michael H. *Arctic Tundra* (Habitats). Danbury, CT: Children's Press, 1997.

George, Jean Craighead. *1 Day in the Tropical Rain Forest* (Newbery Medal Winner Series, No. 5). New York: Crowell Co., 1990.

Miller. *Race to Save the Planet*. Pacific Grove, Calif.: Brooks/Cole Publishing Company, 2001.

Mongillo, John; Zierdt-Warshaw, Linda. *The Encyclopedia of Environmental Science*. Phoenix, AZ: Oryx Press, 2001.

Morgan, Sally. *Acid Rain* (Earth Watch). New York: F. Watts, 1999.

Pollock, Steve. *The Atlas of Endangered Animals* (Environmental Atlas Series). New York: Checkmark Books, 1993.

Pringle, Laurence P. *Rain of Troubles*. New York: Macmillan; Collier Macmillan, 1988.

Ripple, Jeff. *Manatees and Dugongs of the World*. Stillwater, Minn.: Voyageur Press, 1999.

RECORD KEEPING

The Progress Charts on these pages are for use with questions that follow the stories in the Life Science, Earth-Space Science, Physical Science, and Environmental Science units. Keeping a record of your progress will help you see how well you are doing and where you need to improve. Use the charts in the following way:

After you have checked your answers, look at the first column, headed "Questions Page." Read down the column until you find the row with the page number of the questions you have completed. Put an X through the number of each question in the row that you have answered correctly. Add the number of correct answers, and write your total score in the last column in that row.

After you have done the questions for several stories, check to see which questions you answered correctly. Which ones were incorrect? Is there a pattern? For example, you may find that you have answered most of the literal comprehension questions correctly but that you are having difficulty answering the applied comprehension questions. If so, then it is an area in which you need help.

When you have completed all of the stories in an unit, write the total number of correct answers at the bottom of each column.

PROGRESS CHART FOR EARTH-SPACE SCIENCE UNIT

Questions Page	Comprehension Question Numbers				Total Number Correct per Story
	Science Vocabulary	Literal	Interpretive	Applied	
51	1,2,3	4	5	6,7,8	
53	1	2,3,4	5	6,7,8	
55	1	2,3	4,5	6,7,8	
57	1	2,3,4	5,6,7	8	
59	1	2,3,4	5	6,7,8	
61	1	2,3,4	5	6,7,8	
63	1,2	3	4,5	6,7,8	
65		1,2,3,5	4,6,7	8	
67	1,2	3,4	5,6	7,8	
69	1	2,3	4,5,6,7	8	
71	1	2,3,4	5,6	7,8	
73	1	2,3	4,5	6,7,8	
75	1	2,3,4	5	6,7,8	
77	1	2,3	4,5	6,7,8	
Total Correct by Question Type					

PROGRESS CHART FOR LIFE SCIENCE UNIT

Questions Page	Comprehension Question Numbers				Total Number Correct per Story
	Science Vocabulary	Literal	Interpretive	Applied	
7	1	2,3	4,5,6,7	8	
9	1	2,3	4,5,6	7	
11	1	2,3	4,5	6,7,8	
13	1	2,3,4	5,6	7,8	
15	1	2,3	4,5,6	7,8	
17	1	2,3	4,5,6,7	8	
19	1	2,3,4	5,6,7,8		
21	1	2,3,6	4,5,7,8		
23	1	2,3	4,5	6,7,8	
25	1	2	3,4	5,6,7,8	
27	1	2,3	4,5	6,7,8	
29	1	2,3	4,5	6,7,8	
31	1,2	3,4	5,6,7	8	
33	1	2,3	4,5,6	7,8	
35	1,2,3	4	5,6	7,8	
37	1	2,3,4	5,6	7,8	
39	1	2,3,4	5	6,7,8	
41	1	2,3,4	5,6,7	8	
43	1	2,3	4,5,6	7	
45	1,2	3	4,5,6	7,8	
Total Correct by Question Type					

PROGRESS CHART FOR ENVIRONMENTAL SCIENCE UNIT

Questions Page	Comprehension Question Numbers				Total Number Correct per Story
	Science Vocabulary	Literal	Interpretive	Applied	
111	2	1,,3,4	5	6,7,8	
113	1	2,3,4,5,6		7,8	
115	1,2	3	4,5,6	7,8	
117	1	2,3	4,5,6,7	8	
119	1	2,3	4,5,6,7	8	
121	1	2,3	4,5	6,7	
123	1	2,3,4	5,6	7,8	
125	1,2	3	4,5,6,7	8	
127	1,2,3,4		5,6	7,8	
129	1	2,3	4,5	6,7,8	

Total Correct by Question Type

PROGRESS CHART FOR PHYSICAL SCIENCE UNIT

Questions Page	Comprehension Question Numbers				Total Number Correct per Story
	Science Vocabulary	Literal	Interpretive	Applied	
83	1	2,3	4,5	6,7,8	
85	1	2,3	4,5,6,7		
87	1	2,3	4,5,6,7	8	
89	3	1,2,4,5	6,7		
91	1	2,3,4	5,6	7,8	
93	1	2,3,4	5	6,7,8	
95	1,2	3,4	5,6,7		
97	1,2	3,4,5	6	7,8	
99	1	2,3,4	5,6	7,8	
101	1	2,3	5	6,7,8	
103	1	2,3	4,5,6,7		
105	2	1,3,4,5	6,7	8	

Total Correct by Question Type

METRIC TABLES

This table tells you how to change customary units of measure to metric units of measure. The answers you get will not be exact.

LENGTH

Symbol	When You Know	Multiply by	To Find	Symbol
in	inches	2.5	centimeters	cm
ft	feet	30	centimeters	cm
yd	yards	0.9	meters	m
mi	miles	1.6	kilometers	km

AREA

Symbol	When You Know	Multiply by	To Find	Symbol
in^2	square inches	6.5	square centimeters	cm^2
ft^2	square feet	0.09	square centimeters	cm^2
yd^2	square yards	0.8	square meters	m^2
mi^2	square miles	2.6	square kilometers	km^2
	acres	0.4	hectares	ha

MASS (WEIGHT)

Symbol	When You Know	Multiply by	To Find	Symbol
oz	ounces	28	grams	g
lb	pounds	0.45	kilograms	kg
	short tons (200 lb)	0.9	tonnes	t

VOLUME

Symbol	When You Know	Multiply by	To Find	Symbol
tsp	teaspoons	5	milliliters	mL
Tbsp	tablespoons	15	milliliters	mL
fl oz	fluid ounces	30	milliliters	mL
c	cups	0.24	liters	L
pt	pints	0.47	liters	L
qt	quarts	0.95	liters	L
gal	gallons	3.8	liters	L
ft^3	cubic feet	0.03	cubic meters	m^3
yd^3	cubic yards	0.76	cubic meters	m^3

TEMPERATURE (exact)

Symbol	When You Know	Multiply by	To Find	Symbol
°F	Fahrenheit temperature	5/9 (after subtracting 32)	Celsius temperature	°C

METRIC TABLES

This table tells you how to change metric units of measure to customary units of measure. The answers you get will not be exact.

LENGTH

Symbol	When You Know	Multiply by	To Find	Symbol
mm	millimeters	0.04	inches	in
cm	centimeters	0.4	inches	in
m	meters	3.3	feet	ft
m	meters	1.1	yards	yd
km	kilometers	0.6	miles	mi

AREA

Symbol	When You Know	Multiply by	To Find	Symbol
cm^2	square centimeters	0.16	square inches	in^2
m^2	square meters	1.2	square yards	yd^2
km^2	square kilometers	0.4	square miles	mi^2
ha	hectares (10,000 m^2)	2.5	acres	

MASS (WEIGHT)

Symbol	When You Know	Multiply by	To Find	Symbol
g	grams	0.035	ounces	oz
kg	kilograms	2.2	pounds	lb
t	tonnes (1000 kg)	1.1	short tons	

VOLUME

Symbol	When You Know	Multiply by	To Find	Symbol
mL	milliliters	0.03	fluid ounces	fl oz
L	liters	2.1	pints	pt
L	liters	1.06	quarts	qt
L	liters	0.26	gallons	gal
m^3	cubic meters	35	cubic feet	ft^3
m^3	cubic meters	1.3	cubic yards	yd^3

TEMPERATURE (exact)

Symbol	When You Know	Multiply by	To Find	Symbol
°C	Celsius temperature	9/5 (then add 32)	Fahrenheit temperature	°F